SATELLITES TODAY
THE GUIDE TO SATELLITE TELEVISION

by Frank Baylin

PUBLISHED AND DISTRIBUTED
HOWARD W. SAMS & CO.
A Division of Macmillian
and
UNIVERSAL ELECTRONICS, INC.
4555 Groves Road, Suite 13
Columbus, Ohio 43232, U.S.A.
614-866-4605

Library of Congress Catalog Card Number 84-070697

First edition
     First Printing   March 1984
     Second Printing   April 1984
     Third Printing   September 1984
     Fourth Printing   December 1984
     Fifth Printing   May 1985

ISBN 0-917893-01-8

ACKNOWLEDGEMENTS

Many people provided valuable contributions to the goal of producing an accurate and easily understandable book. Brent Gale, co-author of my second book, taught me a number of invaluable technical lessons which helped to greatly improve this revised edition. Bob Ducaj and Dave Sullivan, both of the Aerowave Corporation, reviewed the first edition manuscript numerous times for technical accuracy. Rosalynd Baylin provided invaluable editorial assistance. Marjorie Allison aided in assembling the photographs and proofing the manuscript. Sincere thanks to Roland Heib, director of the National Cable Television Institute, and his staff for their support and for use of their facilities in preparing the first edition. Bill Frederick graciously contributed a number of the illustrations.

DEDICATION

I dedicate this book to Samuel Baylin
father and loving soul.

INTRODUCTION

HARNESSING THE COSMOS

There is a revolution occurring in the way man communicates that is based on extraordinary developments in satellite technology and computers. These man-made devices have altered everything from business structures to the way our children learn and entertain themselves. In particular, earth stations for receiving audio and video entertainment and information are having a dramatic, global effect in the largest cities and the most sparsely populated rural areas.

In recent years satellite antennas or "dishes" have become a common sight. They are the "eyes" of the communication system that now allows more than one hundred and twenty video channels to be received by a television set anywhere on a continental land mass. These are part of the same revolutionary technology makes it possible for national and even global computer networks to exchange information or for a small radio station to broadcast worldwide.

These earth receiving stations are the first highly visible sign that many consumers have of the changes ahead. In the short life of the satellite communication industry, costs of "earth stations" have been dramatically reduced. Less than 15 years ago purchasing a satellite receiving station could require in excess of a million dollars. Now these "earth stations" are so inexpensive and simplified that even a competent "do-it-yourselfer" can receive signals in his backyard from dozens of satellites orbiting more than 22,000 miles away in space!

Satellite communication devices have uses which extend far beyond the familiar urban backyard, farm or mountain retreat. These technologies now allow businessmen in cities scattered all over the world to be linked together in privacy by television. And a whole new breed of broadcasters are now able to relay entertainment and information via satellite from the largest cities to the most remote corners of our nation and our globe.

TABLE OF CONTENTS

LIST OF TABLES

LIST OF FIGURES

THE PIONEERS

Those who benefit from satellite communication systems are in a situation unique in the history of mankind. Until a few short years ago television programming and methods for distributing this entertainment and information were extremely limited. Today the old model of "TV viewing" is being revolutionized by our advancing technology. And these amazing technical developments have occurred in an incredibly short time.

All modern communication is based upon electromagnetic waves which were first detected and measured less than 80 years before the launching of the first communication satellite. In 1888 the German physicist Heinrich Hertz demonstrated that the lengths of radio wave vibrations, known as their wavelengths, could vary in size from centimeters to meters. He discovered that radio waves, which are one type of electromagnetic waves and which are identical to the phenomena called light, have a velocity of 186,000 miles per second and exhibit similar properties independent of their wavelength. This discovery paved the way for the telecommunication explosion of the 20th century.

In 1896 the first wireless radio signals were successfully transmitted and received. Guglielmo Marconi, an Italian scientist, transmitted these signals 9 miles, an amazing distance at that time. The messages were a combination of long and short "clicks" that were played over a loudspeaker and that were then translated into either dots or dashes, as in the familiar Morse Code. Five years later Marconi flashed a radio wave message an even longer distance from England to Newfoundland.

But scientists were not the only ones interested in communicating via the new-found technology. Amateur radio operators, known then and also today as "hams," were allocated the short-wavelength signals since the longer ones had already been claimed by the pioneers and scientists. Not long after that in 1923, a 100-meter, short-wave message was transmitted transatlantically by such hams. In 1900 Reginald Fessenden, a Canadian, first relayed a poorly articulated human voice over one mile via radio waves. Then in 1906 he successfully radioed a distance of 200 miles to New York state and was unintentionally overheard in Scotland. This was an astounding feat at the time.

These experimental developments led in 1919 to the world's first commercial radio station, CFCF in Montreal, Canada. KDKA, the first American station had its premier broadcast on November 2, 1920. Before long, hundreds of radio stations were springing up all over the world.

For the first time in the history of man, the public could be reached instantly. This revolutionary change in communication and in social structure inspired enthusiastic experimentation and invention. Short-wave (SW) broadcasts, which had been developed by ham radio operators, were soon shown to travel around the world in less than a second. These were then used by both hams and private and public industries. Commercial radio broadcasters, were allocated the relatively shorter AM wavelengths. In 1935 Edwin Armstrong, an advocate of ultra short wavelength communication, demonstrated a new broadcasting system, FM, which was an improvement over AM radio. Derived from the abbreviation for "frequency modulation," FM is relatively free of atmospheric disturbances and is characterized by its excellent fidelity.

The next important development in the use of radio waves was for relaying images or pictures over-the-air. In 1904 Arthur Korn, a German, first sent "wire-photos" 600 miles via a telephone line from Munich to Nuremberg. His transmission of wire photos from the European continent to England in 1907 caused a sensation. The adage "A picture is worth a thousand words" could not have been more true than at this time of new discovery.

In 1922 Korn demonstrated "phototelegraphy," in which a wireless receiver was hooked to a typewriter that printed various sized dots. In this manner, a picture could be "painted" in 40 minutes. The first transatlantic radio broadcast of pictures from London to New York was accomplished in 1924 by Richard Ranger, an American, using a more sophisticated device. His mechanism utilized a spot of light that scanned a picture. Reflected light was picked up by a photocell that converted the different intensities into electrical signals. These were then relayed by radio to a receiver that changed them back into light. The light was projected onto film, which was subsequently developed and printed into photographs.

These pioneering advances were the forerunners of modern communication techniques.

Television, a fixture that today seems such a familiar and natural part of our world, is based upon the work of many scientists and experimenters. The very first references to television came in 1910 in the Kansas City Star. A story entitled "Television on the Way" described experiments by a French scientist. Some time passed before C. Francis Jenkins, an American TV pioneer, predicted in 1922 that "motion pictures by radio in the home and an entire opera some day may be shown without the hindrance of muddy roads."

In 1884 Paul Nipkow, a German, invented the Nipkow disk, which dissected an image into fragments of light for detection and subsequent broadcast. In order for these fragments to be reassembled into an image, it was necessary to invent a device to recreate the picture. The step that paved the way for such a breakthrough was the invention of the cathode-ray tube by Sir William Crookes, a British scientist, in spite of the fact that he considered his device of 1895 merely a scientific "plaything." The modern picture tube, developed by Vladimir Zworykin, is a sophistication of Crookes' "plaything." The Crookes tube used electron beams that illuminated a screen, but Zworykin was the first to demonstrate control of this electron beam solely by electronic signals. As a result, with the perfection of this technology, television could officially be introduced to the public at the 1939 New York World's Fair.

At the same time, scientists were developing another revolutionary and important technology, radar, which reached its peak of development during World War II. The concept underlying radar is quite simple. Radio waves are broadcasted toward an object, and a receiving antenna then scans for any reflected waves. This antenna is capable of detecting any aircraft, bird, or other reasonably sized object in its path in fractions of a second. During World War II the Congressional Record referred to radar as "the superweapon, the most revolutionary military device of this war." Radar was critical in the defense of Britain because a 24-hour watch of the North Sea coast was maintained. When the war was over more peaceful uses were found for radar, as is so often the case with advances made during a war.

The techniques underlying radar have been used in "line-of-sight" television relays throughout the world. Because TV signals have to be sent between antennas in view of each other to compensate for the earth's curvature, stations have to be installed along the way at intervals of approximately 30 miles. However, these relay stations were and still are too expensive for use in broadcasting television to small, remote communities. The answer to this problem was developed by the cable TV industry.

Cable TV companies in rural areas erected master antennas capable of detecting signals broadcast from the nearest large city. The master antennas were similar to ordinary rooftop TV antennas but were much more powerful, e.g. often larger in size or of better quality. The signals would then be relayed via cables into households that pay a fee for this service. Transmission cables could be buried underground, strung alongside telephone wires, or in a few instances laid on top of the ground. In this way, people in remote areas were served with the same information that their urban neighbors received.

The most pervasive and powerful communication technology, satellite broadcasting, stems from all the preceding developments. The history of satellite communication is filled with both dreamers and realists whose efforts have mated fantasy with practical science and engineering to achieve the impossible. In 1911 H. Gernsback suggested the use of radio, television, and radar in his science fiction book, *Ralph 124C41*. In 1942 George O. Smith wrote a story entitled "Venus Equilateral" about a manned space vehicle which orbited the sun and which was designed specifically for communication. However, the most fascinating prediction of the future was Arthur C. Clarke's article, "Extra Terrestrial Relays," which appeared in the October, 1945 edition of Wireless World. This work in describing a complete, integrated worldwide communication network of communication satellites had a revolutionary impact.

Satellite communication at first involved massive, expensive, and often unreliable equipment. Communication using microwaves in space was first accomplished in 1948 by the U.S. Army Signal Corp's Project Diana. By bouncing radar signals off the moon and back to earth again, they proved that relatively low power could be used to transmit and detect signals over extremely long ranges. In 1954 the U.S. Navy successfully reflected the first voice messages from the earth to the moon and back again. Within two years of this event, a U.S. Navy moon relay service was established between Washington, D.C. and Hawaii. Such long-distance communication was dependent only on the presence of the moon, with signals taking only a few seconds to travel to the moon and back.

In the late 1950's the activity accelerated with the launching of the world's first satellites. On October 4, 1957, the USSR successfully rocketed Sputnik I into orbit. On New Year's Day, 1958, the U.S followed suit by launching its first satellite, Explorer I.

In 1958 the U.S. Congress established the National Aeronautics and Space Administration (NASA) in order to foster satellite communications and space exploration. The first NASA-sponsored satellite was SCORE, launched by the U.S. Air Force. SCORE was an important advance being an active relay. It received messages at 150 million cycles per second, taped them, and then relayed the information back to earth at a different frequency of 122 million cycles per second. On December 19, 1958, President Eisenhower recorded and transmitted the very first satellite message, a Christmas greeting. SCORE survived a brief lifetime of just 12 days.

In 1959 the moon was used as a passive reflector for signals sent from the Bell Telephone Laboratories in Holmdel, N.J. A live voice was transmitted from the Bell Labs to the moon and then back to the Jet Propulsion Laboratories in Goldstone, California. This was the first of 17 tests of Project Moonbounce, which all used the moon as a mirror for radio waves.

The next major project managed by these research centers (the Bell Telephone and the Jet Propulsion Labs) was the construction of ground communication equipment for Echo I, a passive satellite that simply functioned as an orbiting reflector. It was a 100-foot-high balloon made of Mylar only 5/10,000 of an inch thick which was launched on August 12, 1960. The signal reflected back to earth was only a millionth of a millionth of the original 10,000-watt uplinked transmission. This experiment demonstrated that two-way telephone conversations across the United States by satellite were possible even when signals had extremely weak powers.

The pace of these developments continued unabated. To test the feasibility of transoceanic communication via satellite, RCA designed and built the Relay satellite in 1961. This new spacecraft had improved abilities to transmit telephone, telegraph, and television messages across the ocean. In 1961 the Hughes Aircraft Corporation built Syncom, an experimental, active satellite that was to be launched into a 22,300 mile high geosynchronous orbit. In such an orbit Syncom's velocity matched the speed at which the earth rotated and therefore would be stationary relative to any observer on the earth below. In contrast, previous satellites that had been placed in elliptical orbits were visible from any one location on the earth for only limited periods of time.

Project Telstar, designed and built by AT&T and its subsidiary the Bell Telephone Laboratories, linked Europe and North America via television on July 10, 1962. This launched an explosion of activity that molded what we now know as the "Information Age". Telstar was an active satellite because it contained a device called a "transponder" to receive, amplify, and retransmit signals to earth.

Although cosmic radiation destroyed Telstar's electronics only 226 days after launch, it paved the way for the whirlwind advance of domestic and international satellite communication.

Relay I, launched by NASA on December 13, 1962, was an active satellite similar in function and operation to Telstar. It was the first spacecraft to link North and South America. On February 14, 1963, the first Syncom satellite was launched by NASA for Hughes Aircraft. An unsuccessful geosynchronous orbit was attempted; the on-board communication system did not operate. Only a few months after the failure of Syncom I, Telstar II, improved substantially over the earlier model because it was resistant to the radiation that put its predecessor out of commission, was launched into orbit for the Bell System by NASA. On July 26, 1963, two months after Telstar II, Syncom II was successfully placed into a geosynchronous orbit above the mouth of the Amazon River.

The next satellite of this series, Syncom III, relayed the 1964 Olympic Games in Tokyo to the United States and the 1964 World Series to Japan. On April 6, 1965, the first international satellite, Intelsat I known as Early Bird, owned by the Intelsat Organization and built by Hughes Aircraft Company, was launched. Early Bird, so named because of its position in the far eastern sky, linked viewers and programmers in North America and Europe. Intelsat (the International Telecommunication Satellite Consortium) had been founded in 1964 to own and operate the "free-world's" international satellite communication system.

These pioneer vehicles have paved the way to the launching of a myriad of increasingly sophisticated satellites. This had included more than 20 Intelsat vehicles and 12 Soviet Molniya satellites by 1969, as well as more recent Soviet satellites Statsioner, Reduga, and Ekran. Many countries today have communication satellites. For example, Indonesia has launched Palapa, Canada has the Anik, and Japan has its Sakura series of spacecraft for domestic communication.

During the 1970's, important experiments were underway that would provide information necessary for the development of an advanced and successful commercial satellite communication system. One such experiment was the Communications Technology Satellite (CTS) Program, a joint U.S./Canadian venture. It utilized the Canadian Anik series of vehicles, which were sent messages from earth to be retransmitted to 36 remote receiving stations. This ATS-65 satellite was used in experiments in both India and the Rocky Mountains. These proved that a spot beam from outer space covering 170,000 square miles could be received using only $2000

Figure 1-1. Syncom II. The world's first successful synchronous satellite, was launched from Cape Canaveral on July 26, 1963 to open a new era in communication. After completing initial experiments above the Atlantic Ocean relaying messages between Western Europe, Africa and the United States, the spacecraft was employed in a public demonstration at the 1963 ITU meeting in Geneva. At this time delegates exchanged trans-Atlantic phone greetings with United Nations and U.S. government officials in New York and Washington. Later, the satellite was "walked" half-way around the world to a position above the Indian Ocean where it remained operational for over four years for the Department of Defense. Syncom II was decommissioned and retired from service in April 1969, more than five years after its expected life span. *(Courtesy of Hughes Aircraft Company)*.

worth of electronics. At least twenty of these powerful beams would have been needed to service all of the continental United States.

Today, the experiments and the ensuing progress continue. More powerful and sophisticated satellites for broadcasting and for other special uses are constantly being developed as the basic technology is improved.

Social and Economic Developments

These rapid advances in technology were accomplished by formation of new government institutions, laws, and businesses. It is interesting to realize that many present-day issues and problems in satellite communication had also been confronted earlier in the development of radio and television.

In the pioneer days of radio as is the case today, the production and broadcasting of programming had to be paid for in some fashion. Proposals for doing so ranged from attaching a coin box on all radios to transmitting programs over rented telephone lines. During this debate advocates of free radio argued for freedom of the airwaves. AT&T finally solved the problem by offering time for sale to commercial sponsors. The rapid growth in the number of radio stations from 3 to 595 between January, 1922, and January, 1923 attested to the feasibility of the technology and the economic potential of advertising. Finally, in 1926, with the million-dollar purchase of station WEAF, the RCA network was formed. Radio had come of age.

In the same year, the Federal Radio Commission (FRC), predecessor of the Federal Communications Commission (FCC) was formed. The goal of this government body was to "make available, so far as possible, to all people of the United States, a rapid, efficient, nationwide and worldwide wire and radio communications service with adequate facilities at reasonable charges." Their regulation consisted almost entirely of issuing licenses on a "first-come, first-served" basis.

Years later, the FCC played a more influential role in the development of television broadcasting. While cable TV companies were slowly bringing service to rural areas, the FCC was attempting to foster the use of UHF (ultra high frequency) channels in these isolated areas. However, UHF channels had characteristically poorer quality transmission than did the standard VHF (very high frequency, channels 2-13) broadcasts. The material being aired was also lower in budget and scope. As a result, cable TV became the favored distribution method by many consumers.

Figure 1-2. Early Bird. Intelsat I known as Early Bird, the world's first fully operational communication satellite, was designed and built by Hughes Aircraft for Intelsat under management by Comsat. Here it gets final adjustments in preparation for space-simulated vacuum tests at Hughes Aircraft in Los Angeles. The spacecraft was launched on April 6, 1965 into a geosynchronous orbit 22,300 miles over the Atlantic to provide 240 two-way telephone channels or one two-way television channel between Europe and North America. The satellite could also simultaneously handle teletype and facsimile communications and telephone conversations. It has been replaced by more advanced Intelsats. *(Courtesy of Hughes Aircraft Company)*.

As cable companies realized profits in the endeavor of relaying already existing programming to remote areas, they began entering a new, profitable market, the highly populated urban centers. In the early 1960's cable companies began expanding into the business of purchasing, reselling, and often producing programming.

While cable television was growing, the technical and regulatory groundwork was being laid for the satellite telecommunication industry. In 1961 the United Nations adopted a resolution for the peaceful uses of outer space in the interest of world harmony. In 1962 President Kennedy signed the Communication Satellite Act to encourage and support growth in this area. Comsat, a private corporation established to plan, own, and operate a commercial satellite communication system and funded and regulated by the U.S. government, was given the sole rights to American participation in the launching and ownership of satellites for international use.

By 1965 the technical and economic developments in satellite communication prompted a number of large American corporations to enter the field without government financial aid or the support of international institutions. ABC realized that a substantial savings of $25 million per year could be achieved through the use of satellites in order to escape from being tied to terrestrial video circuits of AT&T. This network then requested permission from the FCC to operate its own communication satellite. In response and in the interest of fairness, the FCC asked for proposals from other parties in the private sector interested in developing a commercial satellite system for the continental United States. RCA Global Communications suggested launching a system of three vehicles that would serve 300 earth-based receiving stations, while CBS proposed a system for use by all the networks that would also carry PBS programs at no charge. Comsat responded by requesting to build a similar system, which would also provide programming for cable TV stations and information for the press wire services. Teleprompter, Western Tele-Communications, AT&T, Hughes Aircraft/GT&E, Western Union, Fairchild Space and Electronics, and MCI/Lockheed/Comsat Satellite also expressed similar interests. Surprisingly, all estimates for such systems were within 20% of each other.

Because of the similarity in costs, the FCC was in a position to make a decision based on other factors. RCA and Western Union were chosen because they had technical expertise and sound business plans. Although these two companies were first in the "space-segment" business, many other companies have since joined in building, launching, and operating satellites for the many different modes of communication.

The cable TV industry began its participation in the satellite communication business at the beginning of the 1970's. Pioneers such as Irving Kahn, president of Teleprompter, realized that cable TV could offer greatly improved programming and service by using satellites. In June, 1973, the Canadian Anik satellite was launched by NASA and leased by RCA American Communications to become the first satellite to serve the American domestic market. Anik was followed only one year later by the launching of Western Union's satellites, Westar I and II. On September 30, 1975, Home Box Office, a network catering to cable television, inaugurated pay TV services by broadcasting the Ali-Frazier boxing match, the "Thrilla in Manilla." To achieve the necessary link for this landmark production, RCA leased time on the Westar I vehicle. By the end of 1976 there were 130 earth receiving stations serving cable companies. By the 1980's every cable TV station in the country had one and usually two or more satellite receiving antennas as part of their sophisticated and essential earth stations.

Earth stations that only receive signals from outer space are not the unattainable commodities they once were. For example, in 1967/68 the first stations purchased by Comsat had cost one quarter of a million dollars each. In 1976 the FCC approved the use of the smaller 4.5-meter (15 feet) dishes, and prices rapidly dropped from $75,000 to $25,000 for an antenna, a low noise amplifier, and two receivers. By 1986, consumers could purchase quality home satellite systems ranging in price from $1000 to $3000!

The number of TVROs sold and installed mushroomed as the costs plummeted. By the end of 1982 there were, by conservative estimates, 45,000 home earth receiving stations installed. By comparison, in 1980 and 1981 there were 5,000 and 15,000 private terminals sold. It has been estimated that 180,000 and 600,000 home systems were installed in 1983 and 1984, respectively. By the end of 1985 an estimated 1.7 million systems were receiving satellite television. The industry had come of age.

This continuing rapid drop in prices and increasing numbers of earth stations has taken us back full circle to issues similar to those faced in the early days of radio. We have squarely confronted the question of who pays the program producers. In addition, the number of networks and programmers has dramatically increased and, as a result, regulators such as the FCC and Congress are under pressure to quickly find solutions to difficult problems. And the rapid introduction of new technologies has made resolution of these issues even more complex and necessary.

REACHING FOR THE STARS—THE UPLINK

Every satellite communication circuit is composed of an uplink, a telecommunication satellite, and an earth receiving station. The uplink beams messages in the form of radio waves into outer space just like a headlight of a car illuminates a dark road. Satellites receive these weak signals, amplify or boost their power, and then retransmit them back to earth. Earth stations, commonly known as TVRO (television receive-only) stations, detect the very weak signal received from space and extract the original message much like uplinks working in reverse.

TVRO earth stations operate at substantially lower power levels than do uplinks, which must send signals powerful enough to reach a satellite located more than 22,300 miles away. Therefore, while both uplinks and receiving stations may be located anywhere "in view" of a satellite, uplinks are much more expensive and are more carefully regulated by the FCC. It is fortunate that homeowners and TV station operators can use the less expensive earth receiving stations and that transmissions from each sophisticated uplink can be detected by an unlimited number of TVROs.

Uplinks serve many segments of the business community including TV and radio stations, telephone companies, as well as other large organizations. In many cases, an on-site uplink is fed by direct cable links from broadcast studios, telephone lines, or computers. Many television stations relay signals by conventional, over-the-air methods to distant uplinks, which then send their signals to communication satellites. Numerous organizations also need uplink facilities on an occasional basis for teleconferencing, sporting events, or other special happenings. In such cases, the option of "taking Mohammed to the mountain," i.e., bringing the relay equipment directly to the action, is an economical alternative to broadcasting over-the-air to a fixed uplink site. In addition, the leasing of portable

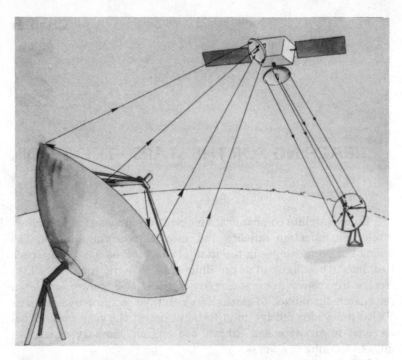

Figure 2-1. A Satellite Communication Circuit. Broadcast signals are relayed by an uplink via a geosynchronous satellite to many TVROs.

Figure 2-2. A Portable Uplink. This 3-piece, 5-meter antenna on a hydraulic mount can be set up to transmit to a communication satellite in two hours. All the electronic equipment is situated inside the 34 by 8-foot wide trailer. *(Courtesy of the Public Service Satellite Consortium, PSSC).*

uplinks complete with all equipment including operating crews, is a viable option since owning a complete uplink is expensive with costs ranging from $250,000 to $660,000.

Whatever option is used to uplink signals to communication satellites, programmers of all bents are now capable of broadcasting to national or even international audiences at affordable prices. The great power of this technology lies in its ability to serve even those located in the most remote areas.

The natural phenomena that ties these programmers and audiences together and that links business information networks is the electromagnetic wave. The equipment that allows us to use electromagnetic waves is one of the great triumphs of modern technology.

What are Electromagnetic Waves?

Electromagnetic waves are the agent by which radio, television, satellite communication networks and many other man-made devices work. What are these invisible but extremely useful phenomena of nature? If we could see an electromagnetic wave, it could look like the waves that travel outward in concentric circles when a pebble is tossed into a pond. Electromagnetic waves are also similar to the waves of vibrating air molecules that we know as sound, which travel through the air at 760 miles per hour. Electromagnetic waves, vibrations of electric and magnetic fields, travel much faster than sound at 186,000 miles per second, the speed of light. At this speed, an electromagnetic wave travels back and forth between the earth and a satellite almost instantly in approximately half a second.

One of the most important properties of electromagnetic waves is their frequency, the number of vibrations that occur every second (see Figure 2-3). Just as the frequency of sound vibrations determines whether a musical note is either a soprano or a bass, so the frequency of electromagnetic waves determines whether the vibrations may be in forms such as X-rays or visible light. Electromagnetic waves sent from earth to satellites usually have frequencies of approximately 6 billion cycles per second, abbreviated as 6 gigahertz.

A second important characteristic of electromagnetic waves is their wavelength, the distance in which one complete vibration occurs. Since electromagnetic waves all travel at the same speed, one that has a shorter wavelength must have a higher frequency, and vice versa (see Figure 2-3).

Electromagnetic waves are also defined by their polarization (see Figure 2-4). To understand this idea we can imagine a car driv-

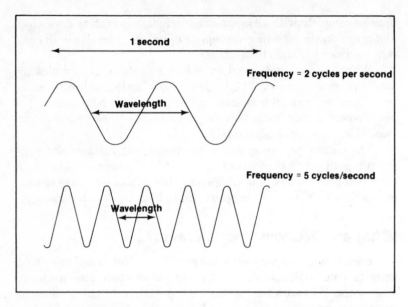

Figure 2-3. Electromagnetic Waves. This schematic representation of electromagnetic waves shows that when the wave length is longer the
frequency is lower and vice versa. For example, a 4 gigahertz or 4
billion cycle per second frequency has a 7.5 centimeter wavelength while a 2 GHz signal has 15 centimeter wavelength.

ing along a highway. It can travel to the same destination by following a curving road along flat ground or by following a straight road
over rolling hills. These types of movements are classified as either
horizontal or vertical polarization.

Electromagnetic waves having the same frequency and wavelength can have different powers. If the power is high, precautions
must be taken to prevent damage to people and property. For example, the strength of microwaves (one form of electromagnetic waves)
near uplink antennas is high enough to cause serious burning or
long-term damage such as eye cataracts. Therefore, the uplink area
must be securely fenced in and protected. In cases where uplinks are
not manned, the FCC requires that they be operated by remote control so that the danger of beam misalignment is reduced. There is no
such problem with earth receiving stations at the other end of the
satellite communication circuit, since they operate with low powers
comparable to those of a stereo system.

Many seemingly different phenomena encountered in nature
including light, X rays, infrared heat rays, microwaves used in communication, and gamma rays from the cosmos are all electromag-

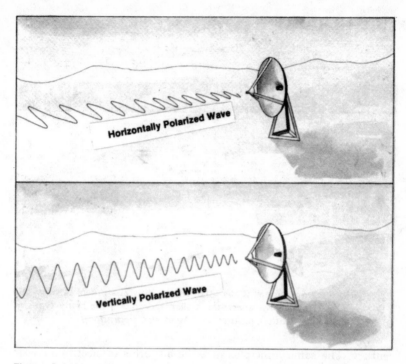

Figure 2-4. Vertically and Horizontally Polarized Electromagnetic Waves. Waves of these two polarities vibrate at right angle to each other.

netic waves (see Figure 2-5). Surprisingly, the only difference among them is their frequency. For example, the wavelength of visible light is comparable to the dimensions of atoms and molecules, and their frequencies are on the order of millions of billion of cycles per second. In contrast, microwaves have lower frequencies of billions of cycles per second and wavelengths that range from meters to centimeters. Lower frequency waves, called radio waves, used in conventional radio transmissions have frequencies of millions of cycles per second and wavelengths which can be miles long.

Modern Methods of Communication

Until recently, man has been blind to all but those forms of electromagnetic waves known as light and infrared heat rays. As technology has improved, so has our ability to create and detect such waves for use in communication. The earlier simplest form of man-made radio communication, Morse Code, has very rapidly evolved to radio, television, and now to satellite communication in less than a

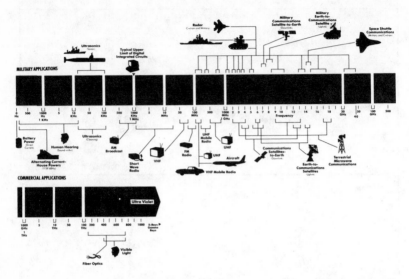

Figure 2-5. Electromagnetic Wave Allocations. This chart depicts how the electromagnetic spectrum is allocated among the many types of communication. *(Courtesy of Avantek Corporation).*

century. The same principles underlie all forms of electromagnetic wave communications.

Analog and Digital Signals—Coding the Message

Any message, whether it be the image and voice of an entertainer or details of stock-market transactions, must first be changed into a form that can be superimposed onto radio waves. If an analog coding method is used, the pattern of a message is mimicked by electrical voltages. For example, a voice can be changed into an analog signal by a microphone that creates a voltage determined by the loudness of the sound. The louder the sound, the higher the voltage.

A digital coding method uses the numbers 0 and 1 to convey all information. For example, a voice could be expressed in digital form if the loudness at each point in time was expressed by numbers written as patterns of 0's and 1's. A photograph can also be described by a long series of 1's and 0's that are coded so that some impart information about the location of the dots composing the picture and others determine the brightness and color of the dots. Computers exclusively use digitally coded information.

Uplinks can relay either digital or analog forms of the same message. Converters that can translate between these two languages are available. For example, conversations between computers relayed by satellite are always digital, while most TV broadcasts are expressed in analog form. However, the trend is towards the use of digital broadcasts as newer and higher-quality video transmission codes are being developed and as satellites are being packed with more sophisticated electronics which allow higher amounts of information to be transmitted.

Modulation—Adding the Message to Radio Waves

Analog or digital signals are added onto electromagnetic waves by a process called "modulation." Once the message is modulated onto an electromagnetic wave which is known as the carrier wave because it carries the information, it can be relayed from a sending to a receiving antenna. Radios, televisions, and other communication equipment demodulate the signals they receive; ie., they extract the original message from the carrier wave (see Figure 2-6).

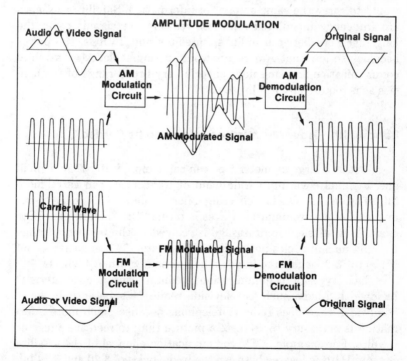

Figure 2-6. Amplitude and Frequency Modulation of Signals. Two principle methods for superimposing audio, video or data messages onto carrier waves are amplitude (AM) or frequency modulation (FM). These resultant signals can then be transmitted by cables, over-the-air, by fiber optic lines or via satellites.

The simplest method to modulate a carrier wave is to switch it on and off. For example, Morse Code is relayed as series of dots and dashes by turning the carrier wave on and off. The most familiar methods of modulation used today are amplitude modulation (AM) and frequency modulation (FM) as encountered, for example, in AM and FM radio. In amplitude modulation, the power of a carrier wave is varied in accordance with the message being relayed, while in frequency modulation the frequency of the carrier wave is varied (see Figure 2-5).

Each type of modulation has advantages and disadvantages. AM messages must have relatively high powers to be able to travel long distances without being weakened so severely as to impair clear reception. They are also more prone to picking up static than are FM messages. On the other hand, FM signals need relatively lower powers for successful, long-distance transmission, but must use a substantially wider range of frequencies than AM messages would to carry the same amount of information. Satellite messages are frequency modulated for these reasons. As signals cover the long distances between uplinks, satellites and TVROs, the power becomes so low that AM relays would be unusable. Also, satellite communication transmissions cover a very broad range of frequencies as is required by FM broadcasts.

Bandwidth—How Much Information Can be Carried?

Just as a large-diameter pipe can carry more water than a small one, a signal covering a wide band of frequencies can carry more information than can one covering a narrow band. This range of frequencies of electromagnetic waves is termed the "bandwidth." For example, if a television message is relayed in the frequency range from 54 million cycles per second (abbreviated 54 megahertz, or 54 MHz) to 58.2 megahertz, it would have a bandwidth of 4.2 MHz.

Each type of communication medium uses a characteristic bandwidth. Media such as television require a substantially wider bandwidth than does radio or telephone because much more information is necessary to recreate a picture than to recreate music or a voice. For example, a TV communication channel of the satellite Satcom III-R is located between the frequencies of 3.70 and 3.74 billion cycles per second, abbreviated as gigahertz, GHz. This channel therefore has a bandwidth of 40 MHz. Voice channels, however normally require a bandwidth of only 3,000 to 4,000 cycles per second for quality sound reproduction.

Amplification—Increasing the Signal Strength

Messages beamed into space from an uplink antenna are weakened on their voyage to a satellite as the signal spreads out and is absorbed by water vapor, clouds, and other atmospheric materials. This weakening is called "attenuation." Amplification, the opposite process, increases the signal strength. In the same fashion that a photograph is enlarged but not changed, amplification retains the original message. All televisions, radios, stereos, and other communication equipment amplify a signal before the carrier waves are demodulated.

Noise—Hindering Clear Communication

In a perfect communication system, signals would be relayed with no interference or noise. However, television or radio broadcasts are occasionally of poor quality or "noisy." Noise is present in all matter at temperatures above absolute zero, the temperature at which all molecular motion ceases. (It is fascinating to realize that there is no temperature colder than absolute zero). Noise is caused by the endless motion of the molecules that compose all matter. These small, vibrating charged particles generate electromagnetic waves that mask the organized signal sent by man. Noise from the environment becomes stronger as the temperature increases. Note that noise is also generated by internal heat in amplifiers, receivers, and other electronic equipment.

The noise that antennas detect also increase as the signal bandwidth increases. So wideband transmissions which characterize satellite broadcasts are often accompanied by substantial amounts of noise.

The quality of a communication circuit is determined by the ratio of signal-to-noise power (see Figure 2-7). For example, if a signal of 10 watts were received along with 5 watts of noise, the picture quality will be poorer than if a signal of 10 watts were received with 1 watt of noise. In the case of television, the signal must be at least 63,000 times the accompanying noise in order for a "high-quality" picture to be received.

FCC Radio Wave Allocations:

The Federal Radio Commission (FRC), and later, the Federal Communications Commission (FCC) in concert with the International Telecommunications Union (ITU) have kept order in the air-

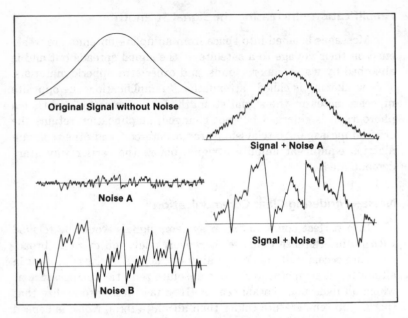

Figure 2-7. The Effect of Noise on a Radio or Television Signal. Noise can distort a signal. If the level of noise, as in case B, is too high the message can be garbled or even unintelligible.

ways by successfully assigning portions of the electromagnetic wave spectrum to different communications media. In fact, the history of man-made communication is unfolded by exploring these frequency assignments (Figure 2-5). As technology was improved, higher-frequency messages were recruited into service. Wire transmission first used relatively low frequencies, since the radio waves used by the pioneers were limited in frequency by the available electronics. When radio waves with frequencies higher than 1.5 MHz were produced by man, the FRC delegated them to "hams" because, at that time, they could not see a use for this region of the spectrum. As technology progressed, coaxial cable transmission, microwave relays, and then satellite communication circuits were allocated successively higher frequencies. However, even today, a relatively small portion of the total possible spectrum is being used.

Table 2-1 shows how a portion of the communication spectrum is being allocated. For example, TV channels 2, 3, and 4 are relayed on carrier waves between 54 and 72 MHz. Channel 2 therefore is relayed on the frequencies ranging from 54 to 60 MHz and has a 6 MHz bandwidth.

TABLE 2-1. FCC ASSIGNMENT OF SOME RADIO FREQUENCIES

Frequency (MHz)	FCC Assignment
3-54	Mobile Radio
54-72	TV VHF Channels 2-4
72-76	Radio Services
76-88	TV VHF Channels 5 & 6
88-108	FM Radio
108-136	Aeronautical
136-144	Government
144-148	Amateur Radio
148-151	Radio Navigation
151-174	Land, Mobile, Maritime, Gvt.
174-216	TV VHF Channels 7-13
216-329	Government
329-890	TV UHF Channels 14-83

Why Microwaves

Microwaves are used in satellite communication for four specific reasons. First, it is important that satellite circuits are used to their maximum potential because the costs of building, launching and maintaining these spacecraft are very high. This requires that wide bandwidths are used. This is possible with very high frequency microwaves because as frequency increases, any given bandwidth becomes a smaller fraction of the operating frequency. For example, a 1 MHz bandwidth located at 10 MHz (100 million cycles per second) occupies relatively more space than does the same bandwidth in the 10 GHz (10 billion cycles per second) region of the spectrum. Since more space is available, wider bands with higher information capacities can be used at microwave frequencies.

A second reason for employing microwaves is based on the need that uplink antennas have to aim a directional beam towards an extremely small target in space. Physics dictates that radio waves can be better focused by an antenna that has a size substantially larger than the carrier wavelength. For example, sending a directional beam of AM radio signals with wavelength of approximately 100 meters would require an extremely large, cumbersome, expensive antenna. Since microwaves of 6 gigahertz have wavelengths of approximately 5 centimeters or 2 inches, a 15-foot uplink antenna can aim most of its radiation into a very narrow beam and relatively low power can be used to reach the target satellite.

Third, microwave transmissions to and from satellites or between line-of-sight stations are not as susceptible to noise from atmospheric disturbances as are lower-frequency transmissions. To illustrate, several times each year for periods as long as two or three days, short-wave radio is useless for long-distance communication because sun-spot activity disturbs the reflection of radio waves by the upper atmosphere.

Fourth, the most important property of microwaves that determines their use in satellite communication is their ability to pass through the upper atmosphere into space. Below frequencies of approximately 30 megahertz, a radio wave beam will be reflected off the ionosphere layer in the atmosphere back towards the earth. Since microwave frequencies are far above the 30-megahertz range, they pass easily through the ionosphere layer.

All these reason suggest exploring the use of ever high frequency microwaves. The trend is in evidence as technology has improved. As a result, the global communication network is constantly being expanded and improved.

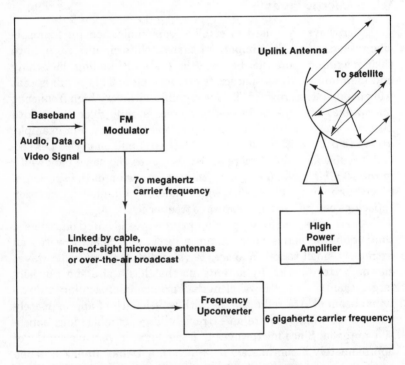

Figure 2-8. Typical Uplink Configuration. Radio, television or computer information are all transmitted to a communication satellite in a similar fashion.

3

OUTPOSTS IN SPACE—SATELLITES

Satellites are a key component in the communications revolution of the late 20th century. During the past twenty years our communication system has been revolutionized because any locations within a satellite's "view" can be linked without the use of expensive cables or line-of-sight relays. Today antennas downlinking video, data and audio information can serve vast areas of our globe at once or can target powerful "spot beams" towards any chosen geographical area.

Communication spacecraft remain in one position relative to any location on earth and can therefore effectively serve as high-volume, full-time switchboards. A satellite in geosynchronous orbit located at 22,247 miles above the equator can accomplish this by rotating at just the same speed as our planet (see Figure 3-1). This is in sharp contrast to the early telecommunication vehicles which, in either low circular or elliptical orbits, were in view from earth stations for only limited periods, had to be tracked by complex, bulky equipment and were able to relay messages for only short durations of time. Each satellite is capable of "seeing" over 40% of the earth's surface and so, in theory, only three "super satellites" would be necessary to link the entire globe (see Figure 3-2). In practice, many more geosynchronous spacecraft, some having "global beam" antennas and others targeting much smaller geographic regions, are rapidly populating the equatorial arc. Today, hundreds of communication satellites encircle the earth like a giant charm bracelet.

Satellite technology is revolutionizing the transmission of information. In particular, the impact on television and radio broadcasting is profound. Before the advent of satellites, broadcasting depended on cables, microwave line-of-sight relays, or the distribution of videotapes to relay programming over long distances. Satellite broadcasting is now by far the most economical delivery method for distances greater than approximately three hundred miles (al-

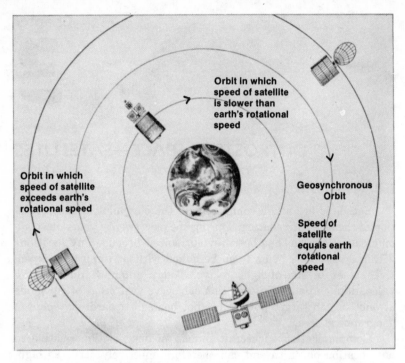

Figure 3-1. The Geosynchronous Orbit. Satellites above or below this special orbit at 22,300 miles over the equator rotate more slowly or rapidly, respectively, than the speed of the earth below. Only those spacecraft at the geosynchronous orbit remain stationary relative to the earth below.

though the widespread introduction of advanced fiber-optic cables will somewhat alter the picture). In addition, since only one "relay station," the satellite, is necessary, many points can simultaneously receive the same broadcast. Also, because a land-based chain of many relay stations are not required to transmit, receive and retransmit a signal, broadcasts can be of much higher fidelity. Another advantage of satellite technology is that radio and television broadcasters need no longer be limited to receive a few programs at a time by a single landline feed, but can now choose from among hundreds of channels of entertainment and information which can be simultaneously recorded for later use.

How do Satellites Work?

A telecommunication satellite which is about the size of a car has the simple function of receiving, processing and rebroadcasting

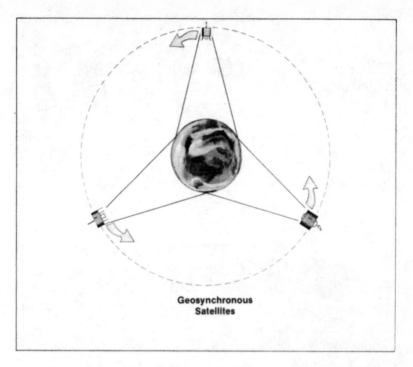

Figure 3-2. A Worldwide Communication System. Arthur C. Clarke predicted in 1945 that only 3 geosynchronous satellites would be necessary to serve the entire globe.

the microwave carrier waves. Signals from uplink stations have very low powers of fractions of a millionth of a watt by the time they reach a satellite. In the heart of a communication vehicle, these microwaves are amplified many thousands of times. Their frequency is then lowered so that the powerful radio waves relayed earthward do not interfere with the original uplinked message. Finally one or more downlink antennas onboard a satellite relay approximately a 5 to 10-watt signal (less than even the lowest power light bulbs) into a well-defined beam to complete the circuit. All the power to carry out these electronic functions is provided by the sun's energy, which is captured by arrays of solar cells. The overall quality of a telecommunication satellite depends both on how accurately receiving and sending antennas are pointing and on the power of the downlinked signal relative to the associated noise.

The number of television channels or other information carried by each satellite is determined by its electronic design. The early Western Union vehicles such as Westar I and II could handle 12

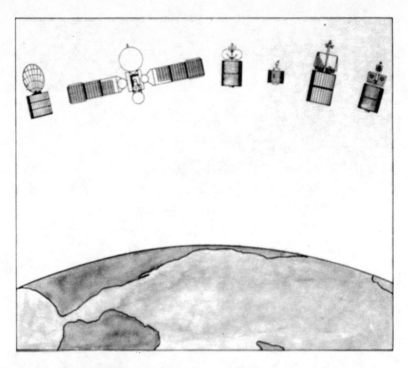

Figure 3-3. Geosynchronous Satellites. Satellites in the geosynchronous orbit encircle our globe in a ring above the equator.

television shows simultaneously; the RCA series of Satcom vehicles transmit 24 channels.

How is the number of channels determined? First, uplinked messages are added to carrier waves in the band of frequencies ranging from 5.925 to 6.425 gigahertz. After the signals are received in space, on-board electronics lowers their frequency to a band spanning 3.7 to 4.2 gigahertz, which still has the same 1/2 gigahertz, or 500 megahertz bandwidth. Engineers who built the early Western Union satellites followed the obvious course of action. They divided this 500-megahertz band into twelve bands of 40-megahertz width which are separated from each other by 4-megahertz-wide protection regions to eliminate "crosstalk" between channels. Each of these twelve channels is relayed by one of twelve circuits called "transponders."

Engineers who designed the Satcom-type spacecraft used a clever technique to squeeze twice as many channels onto each satellite. All the even channels are relayed by horizontally polarized radio waves, while all the odd ones are relayed by vertically polarized sig-

Figure 3-4. SBS Communication Satellite. This spacecraft is shown undergoing a pre-launch test a Hughes Aircraft Company in preparation for its November 15, 1980 boost from Cape Canaveral. The first of a series of vehicles, it was built by Hughes for Satellite Business Systems which is jointly owned by IBM and Aetna Life and Casualty. This satellite provides secure voice, video, teleconferencing, data and electronic mail services to U.S. businesses. It has two concentric cylindrical solar panels, which telescope in space from 9 feet to nearly 22 feet high, to double the spacecraft's solar-power generating capacity over that of many previous communication satellites. *(Courtesy of Hughes Aircraft Company)*.

nals. Thus, transponder 6 has horizontal polarization, and transponder 7 has vertical polarization. In this 24-transponder system each channel, like the 12-channel formatted satellites, has a 36-megahertz bandwidth with 4-megahertz-wide protection region. As shown in Figure 3-7, the centers of the horizontally and vertically polarized bands are offset from each other for further security against crosstalk. Thus, the downlinked signal carries 24 channels, each managed by one of 24 transponders. Any TVRO can select 12 channels from among those having horizontal polarity and 12 channels from the vertically polarized signals.

Figure 3-5. Satcom Satellite. The large wing-like structures are solar cells which provide on-board operational power. *(Courtesy of RCS Americom)*.

Many satellite designs have been developed as the number of vehicles has increased. The accompanying pictures in this chapter give an indication of how many different shapes these spacecraft have. The electronic designs also vary. Satellites are built for many uses and and can have variable numbers of transponders in the C-band (6/4 GHz uplink/downlink frequencies) as well as in other frequency bands. For example, Satellite Business Systems operates the 10-channel, SBS-2 vehicle only in the 14/12 gigahertz mode (Ku-band) in which signals are uplinked in a band from 13.7 to 14.2 gigahertz and downlinked at from 11.7 to 12.2 gigahertz. The Mexican

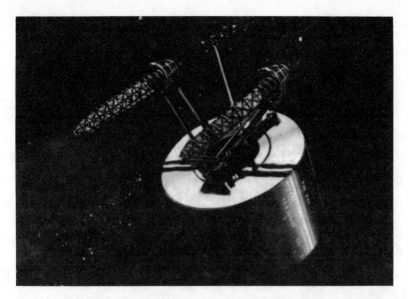

Figure 3-6. LEASAT Satellite. First of the "widebody" satellites, the LEASATs are designed to take maximum advantage of the Space Shuttle's capabilities. The spacecraft are positioned around the earth to relay communications for the U.S. Navy and other military customers. Each satellite is 14 feet in diameter and 20 feet high with antennas deployed. This size will make efficient use of the Space Shuttle's 15-foot payload bay. *(Courtesy of Hughes Aircraft Company)*.

Morelos I and II spacecraft carry a mixture of C-band and Ku-band transponders. The way these transponders are used may also vary. One channel could relay either only one television broadcast or thousands of phone or computer conversations or possibly a mixture of both. Many video broadcasts are often accompanied by "subcarriers" in order to transmit audio information. With the development of more sophisticated digital codes two or more full-bandwidth video signals today can also be relayed over a single satellite channel.

The "footprint" of a satellite describes what signal strength is detected at any geographic region served (Figures 3-9 through 3-12). Footprint maps are provided with every communication satellite to allow earth receiving station designers to size antennas and select low noise amplifiers. For example, the footprint map for Satcom III-R shows lines of equal signal power superimposed upon a map of the continental United States. Footprint maps are labelled with a quantity called the EIRP (the effective isotropic radiated power), a

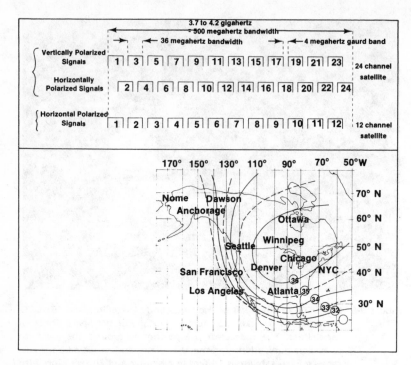

Figure 3-7. Satellite Downlink Frequency Plans. The top portion of this illustration indicates how both 12- and 24-channel satellites are allocated the 500 MHz microwave bandwidth. The bottom illustration
shows a typical footprint map.

common term in the satellite field. EIRP is a measure of power levels reaching the earth. These numbers are expressed in decibels, a
designation often used by communication engineers.

The decibel scale was developed to allow scientist and engineers
to easily describe the enormous changes in power level levels often
encountered. For example, modern amplifiers can easily increase the
strength of a signal by factors of many thousands or million times.
And the signal travelling from a satellite to the earth below will be
attenuated by a factor of thousands of billions. Table 3-1 below
gives examples of how decibels are translated into numbers showing
the relative increase or decrease of power (see Appendix A for more
details). A useful fact to remember is that a small difference in decibels means a relatively much larger change in power levels.

Footprint maps indicating small changes in decibel levels between different geographic locations show how widely signal power
levels do actually change. The signal strength on any footprint map
is always highest right along the direction an antenna is aimed, its

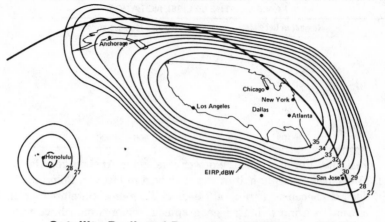

Satellite Radiated Power — 135°W Longitude
Horizontal Polarization

Figure 3-8. Galaxy I Footprint. Those 12 transponders on-board Galaxy I relaying vertically polarized signals provide a maximum of 35 decibels of EIRP. *(Courtesy of Hughes Communications, Inc.).*

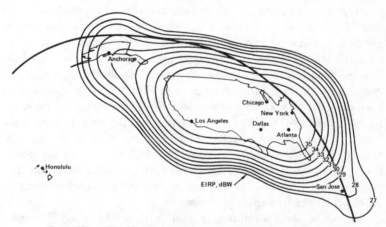

Satellite Radiated Power — 135°W Longitude
Vertical Polarization

Figure 3-9. Galaxy I Footprint. Galaxy I's 12 horizontally polarized transponders include a spot beam on Hawaii in addition to the continental United States in it footprint. *(Courtesy of Hughes Communications).*

TABLE 3-1. THE DECIBEL NOTATION

Number of Decibels	Relative Increase in Power
1	1.26
3	2
5	3.2
10	10
20	100
30	1,000
50	100,000
100	10,000,000,000

"boresight." For example, the power received in Anchorage, Alaska, at 33 decibels is really half of that received in Denver, Colorado, at 36 decibels because there is a 3 decibel difference. Satellites which broadcast to a more limited geographical region would have their footprint covering a smaller area. These spacecraft would have a much higher EIRP than would a satellite broadcasting nationally, because approximately the same power would be more concentrated. Consequently, because footprint maps are determined by the shape and the orientation of a downlink antenna and the power that it transmits, designers of antennas play a very important role.

When a signal leaves the downlink antenna it spreads out in a cone-like beam as it travels to the earth below. This weakening in power is called the "free space path loss." As the distance between an earth station and a telecommunication satellite increases, these free space path losses become even greater. To illustrate, a message would be weaker in Minnesota than in Florida if equal power had been directed to both locations, since Florida is closer to the equator. Also realize that Florida is further off the downlink antenna boresight so it has lower EIRP than a location in, for example, Kansas. It is important to understand that EIRP refers to the power levels at the satellite directed towards each geographic location. The power actually received at any antenna on the earth below must be adjusted for the free space path loss.

Some of the downlinked signal is also lost when it passes through molecules of the atmosphere and is absorbed. Water vapor is the main culprit in the attenuation of downlinked signals. In fact, during a severe rainstorm, the power received on the surface of the earth can be reduced by as much as 10 to 20% for C-band and up to 50% for higher frequency transmissions.

The free space path loss and atmospheric absorption explain why a signal of 9 watts leaving a typical satellite transponder is received on earth at a strength of less than approximately one tenth of a billionth of a billionth of a watt!

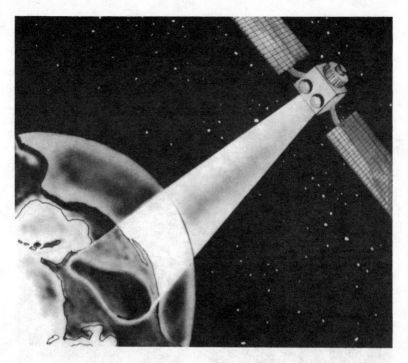

Figure 3-10. A Typical Satellite Footprint. This satellite serves approximately half of the North American continent.

Satellite Launching and Maintenance

The excellence of our satellite communication system depends on launching durable space vehicles into stable geosynchronous orbits. Since the launch of Sputnik I in 1957, man has been able to lift heavier and more sophisticated payloads into orbit. During the 1960's booster rockets were inadequate to position a satellite directly into geosynchronous orbit. As a result, extra rockets in the body of a telecommunication vehicle were used to reposition it from its initial low circular orbit into its final geosynchronous position. The engine and fuel of the satellite accounted for nearly half of its total weight. Today, a similar technique for boosting a satellite into space is used even though current communication vehicles have greater masses and can be lifted into space by more powerful craft. The earlier Thor-Delta launch rockets have been replaced by stronger boosters such as the Delta-2914, the Delta-3912, and the Atlas Centaur. For example, on September 24, 1981, NASA launched the SBS-2 (Satellite Business Systems-2) into a circular

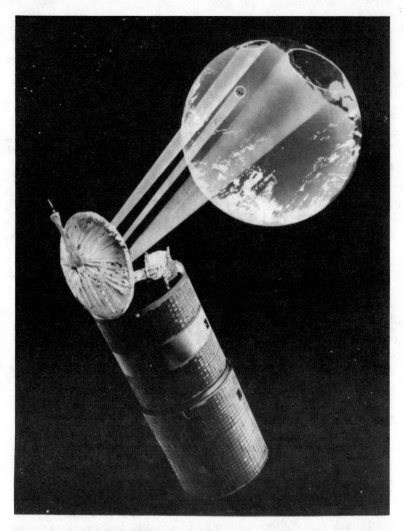

Figure 3-11. Galaxy Satellite. The three Galaxy satellites, built by Hughes Aircraft Company, offer communication services to business and broadcast users. Each vehicle has 24 C-band transponders all of which are available for purchase over the lifetime of the spacecraft. (Courtesy of Hughes Aircraft Company).

orbit at 19,236 nautical miles. This satellite then drifted into its correct orbital slot at a rate of 2.87 degrees east per day, and finally it was boosted into its slightly higher geosynchronous orbit.

Many satellites are now carried into preliminary position by the Space Shuttle. This has allowed heavier, more capable spacecraft to

Figure 3-12. Westar IV and V Satellites. These three satellite, built by Hughes Aircraft Company, have 24 transponders. The solar array generates 800 watts of power over a ten-year lifetime. *(Courtesy of Hughes Aircraft Company)*.

be recruited into service. This trend began in 1983 with the launching of the Canadian Anik C2 and the Indonesian Palapa B vehicles.

Geosynchronous satellites would remain stationary forever in a chosen location over the equator if there were no extra gravitational forces from the sun and the moon, if solar winds did not sweep past our globe, and if the earth was perfectly round. These forces cause any satellite to drift slowly away from its assigned location. There is, however, a "zero-pull" location at approximately 104.5°W, at which all these forces balance. A spacecraft positioned here will not drift. The Canadian satellite Anik DI is located at 104°W and is the western-hemisphere satellite closest to this point. A second "zero-pull" location is found on the exact opposite side of the globe. All satellites to the west of 104.5°W will drift towards the east, and those located to the east of this point will move westward. Consequently, ground controllers are required to adjust the position of a satellite periodically to counteract the tendency towards drifting. All satellites are equipped with small hydrazine gas charged "thrusters" that are fired whenever necessary.

Satellite antennas and solar cell arrays must also be periodically realigned with their targets. In order to relay a beam having a width of approximately 1.25° towards a chosen location on the earth, the antenna must be accurately pointed within 0.2°. This is very important because most satellites, even when accurately positioned, move in a small figure-eight pattern, usually about 15 miles across. Even this very small movement out of their orbital slot can impair reception of signals on earth. On some satellites the solar cell arrays must also be accurately pointed towards the sun so that they will provide a maximum amount of power. Satellites are therefore aligned by spinning their whole body in order to create a stabilizing gyroscopic effect similar to those forces that keep a rapidly rotating top upright. The improved stability of present-day vehicles permits the use of larger antennas which are capable of targeting very small geographic areas.

Satellites do not live forever. Their life expectancy is determined by how long adequate power and stability can be maintained. The on-board repositioning rockets usually run out of hydrazine gas before a ten year period has expired. Also, solar cells are constantly being bombarded with micrometeorites and cosmic rays that slowly wear them out. Either a 30% reduction in the power of the solar cells or expiration of the hydrazine fuel is taken as a signal to retire a satellite from active duty. When ground controllers put a satellite to rest, it is boosted into a slightly higher, non-geosynchronous orbit where it travels away from other satellites or is allowed to drift slowly to one of the two "zero-pull" points on the equatorial ring. Some "dead" spacecraft are now clustered at such a point 22,300 miles above the International Date Line. This satellite burial ground may remain as a reminder of our technological civilization for future generations of archeologists or even for space travellers.

How Much Do Satellites Cost?

The total cost of a satellite is determined by the expenses of building, launching, and operating each vehicle. In the early to mid 1980's the price for building a telecommunication satellite ranged from $25 to $140 million depending on its complexity. For example, the RCA Corporation earned in excess of $100 million building three G-Star spacecraft for GTE Satellite Corporation and a similar amount constructing three Spacenet satellites for Southern Pacific Communications Company. Alascom, Incorporated purchased RCA Americom's Satcom V for $29 million, while Intelsat paid $74 million to Ford Aerospace for constructing three Intelsat V-A satellites. Hughes Aircraft had contracted to build five sophisticated Intelsat VI vehicles for $700 million at a cost of $140 million per satellite.

Costs to launch a telecommunication vehicle have recently ranged from $16 million to $80 million. This dollar value depends upon the size of the satellite as well as upon the organization chosen to launch it. NASA is competing with Arianspace, a joint venture of several European space companies, which launches vehicles from Kourou, French Guiana. For example, Intelsat has spent about $154 million for orbiting three Intelsat V/V-A vehicles onboard one NASA-Atlas Centaur and two Arianspace Ariane-3 rockets, while Western Union Telegraph Company paid Arianspace $22 million to launch its Westar VI vehicle. And Intelsat V,F-4 was lifted into orbit from Cape Canaveral aboard an Atlas-Centaur booster and transferred into geosynchronous orbit for a total of $76 million.

Because of these high launch fees and the potential for lucrative profits other companies are competing for this business. Space Services of Houston and GCH of Sunnyville, California, have planned to charge between $3 million and $5 million per launch. Although their first attempted orbiting of the 53-foot Percheron Rocket ended with an explosion, a second small, test vehicle, the Conestoga, was successfully launched in 1982. Another private venture, Otrag, has considered Argentina as a possible launch site. A third enterprise, the Satellite Broadcasting Company, a joint effort of British Aerospace and N.M. Rothschild and Sons, Ltd., had planned to construct, launch, and operate direct broadcast television satellites to serve the United Kingdom.

Clearly, building and orbiting a telecommunication satellite is a very costly venture. However, over the past 20 years launching technology has substantially improved so that the cost per pound of orbiting a communication vehicle has dropped dramatically (see

Figure 3-13. Intelsat VI. The Intelsat VI communication satellite stands nearly 39 feet high in space with its antenna unfolded and aft solar panel extended. Built by Hughes Aircraft Company for Intelsat, until 1984 the spacecraft was the most massive ever launched. *(Courtesy of the Hughes Aircraft Company).*

Table 3-2). During the same time, improvements in the capacity and sophistication of electronic equipment have allowed the costs per telecommunication channel to drop even more rapidly (see Table 3-3). The cost per voice channel is calculated by adding the costs for building and launching a satellite, by dividing by the number of voice channels that can be transmitted, and by then adjusting for the expected satellite lifetime. A projected cost for a large-scale domestic communication satellite launched by the Space Shuttle is less than $20 per voice channel. This is an amazing one hundredth of the cost for similar capacity incurred on Intelsat I, Early Bird!

TABLE 3-2. COST PER POUND OF SATELLITE VS. TIME

Satellite	Year of Launch	Cost ($ per pound)
Intelsat I	1965	54,117
Intelsat II	1967	23,958
Intelsat III	1968	18,633
Intelsat IV	1971	10,342
Intelsat V	1980	8,964

Costs of Leasing or Owning Satellite Communication Channels

Before the advent of satellites, services provided by cables and later by line-of-sight microwave relays were expensive. For example, in the 1960's, AT&T charged users of their New York-to-Los Angeles video circuit $2809 per hour. In 1974 this rate was lowered to $1832 per hour when the Western Union Telegraph Company introduced a comparable service via satellite for only $715 per hour. By 1981 the most expensive prime-time rate obtainable was $400 per hour.

Until the late 1970's the demand for transponder time was low and the pioneers had ready access to satellite relays. During 1976 the regular per-channel charge for video transponder time on a full-time, 10 hour per day basis offered by RCA was $60,000 per month, or approximately $200 per hour. In late 1979 time could be leased on one Comstar II transponder for $70,000 per month. By 1981 prices had risen even further. AT&T could lease transponder time on Comstar D-2 for $1,664,700 per year, or almost $139,000 per month. In 1982 the American Hospital Video Network leased a higher-frequency transponder on the STS-2 satellite for relaying medical education and news for a whopping $3.2 million per year, or almost $270,000 per month. Even though the rising demand for transponders over this period of time had caused a sharp increase in costs, the real costs per channel fell by 18 times from 1965 to 1980. During this period the consumer price nearly doubled, but the costs per channel fell by 9 times.

Full-time users of satellite communication circuits have options other than leasing time. In the most extreme cases, an organization will actually purchase and launch a communication vehicle. Owners of satellites have also been selling lifetime ownerships of a complete circuit. Such transponder sales offer the owners a faster recovery of their money in times of increasing costs and high interest rates. Users are also protected by being guaranteed both predictable costs and transponder time. As a result, transponders have been purchased on many satellite including the Galaxy I, Spacenet I, and Satcom IV spacecraft.

By 1985 spare transponder capacity was available given the rapidly increased number of orbiting communication satellites. As a result, the costs of leasing or owning a satellite circuit have leveled off. Experts expect that a transponder shortage will probably develop in years to come as more uses are found for satellite circuits and as this technology becomes an even more integral part of worldwide communications.

Users of Satellite Communication Circuits

Users of satellite communication systems can be grouped into five general categories. The satellite common carriers own and operate their own satellite communication systems. These companies include AT&T, Satellite Business Systems, RCA, Western Union, American Satellite Company, Southern Pacific Communications, and Hughes Aerospace. A second group buys full-time use of one or more circuits from the owners of satellites. Third are those businesses that buy blocks of satellite time for resale to others requiring only part-time use of circuits. Companies catering to broadcast television, such as Wold Communications and Hughes TV, have expanded and now provide business communication services. A fourth group, network coordinators, perform a different function by arranging all the necessary ground equipment and facilities as well as satellite time for one-time events such as teleconferencing. The oldest organization which offers these services, the Public Service Satellite Consortium, has been joined by a host of competitors. The fifth category includes all the systems-hardware vendors who supply complete systems for communicating via satellite. Members of this group include the many hundreds of component-hardware vendors supplying those who wish to piece together their own earth-based communication systems.

Communication Satellite Guide

As a result of a strong demand for transponder time, the FCC has been bombarded with requests for operating domestic communication satellites. In late 1980 this regulatory arm of the American government responded by expanding the arc in which communication relays were allowed to be positioned from 70°W to 135°W to a larger 55°W 143°W. This action indicated the FCC's new policy of allowing the number of orbiting satellites to increase. By late 1983 the FCC had processed requests to launch 53 satellites in addition to the 38 American vehicles already in orbit. Similarly, Intelsat and the International Telecommunications Union have been responsible for managing requests from its member nations, so that by the mid-1980s well over 200 communication spacecraft were in orbit.

Table 3-4 outlines the operating and future domestic communication satellites. Although Satcom III-R, Satcom IV, and Galaxy I are the three key satellites devoted to video broadcasting in the United States, the bulk of the in-orbit communication vehicles are and will be used for relaying other types of information. Also, these tables do not list military and international satellites, whose numbers are too many to include here. A brief sample of such satellites demonstrates the activity in these areas. The U.S. Navy operates five FleetSatCom satellites to interconnect its worldwide fleet, while the international marine communication group, Inmarsat, uses transponders on its own spacecraft positioned over the Indian and Atlantic Ocean for marine communication. In June 1981 both the Indian Apple satellite and the meteorological-survey Meteosat satellite were launched by the Ariane organization. Arabsat, a 22-nation Arabian group, operates a regional satellite communication system carrying voice, telex, television, and other information via Ford-built satellites. The Japanese Sakura CS, CS-2a, and CS-2b vehicles are being used to test higher-frequency transmissions, while the Canadian Anik series of outposts have linked one of the first direct broadcast scrambled television systems for remote locations.

TABLE 3-3. COST PER COMMUNICATION CHANNEL VERSUS TIME		
Satellite	Year of Launch	Cost ($ Per Channel)
Intelsat I	1965	20,000
Intelsat IV	1971	1750
Westar	1974	300
Intelsat V	1979	200

TABLE 3-4. DIRECTORY OF DOMESTIC COMMUNICATION SATELLITES (CANADIAN AND AMERICAN)

Satellite Name	Owner	Launch Date	Orbital Position (°W)	Frequency Band(s)
ASC I	ASC/Hughes	1985	128	C,K2
ASC II	"	1988	83	C,K2
ASC III & IV	"	1987/89	64	C,K2
Anik B	Telesat Canada	12/78	109	C,K2
Anik C1	"	1985	107.5	K2
Anik C2	"	6/82	112.5	K2
Anik C3	"	11/82	117.5	K2
Anik D1	"	8/82	194.5	C
Anik D2	"	11/84	111.5	C
Aurora I	Alascom	10/82	143	C
Aurora II	"	1989	146	C
Comstar D1	ComSat-AT&T/Hughes	5/76	—	C
Comstar D2	"	1976	—	C
Comstar D3	"	1984	76	C
Comstar D4	ComSat/Hughes	12/81	127	C
Comstar K1	"	1988	75	K2
Comstar K2	Comsat General/Hughes	1989	130	K2
Cygnus I	Cygnus	1988	43	K
Cygnus II	"	1988	45	K
Fedex A	Federal Express	N/A	77	K2
Fedex B	"	1988	124	K2
Finansat I		N/A	178	C
Finansat II		N/A	48	C
Fordsat F1	Ford Aerospace	1987	101	C,K2
Fordsat F2	"	1987	93	C,K2
GStar A1	GTE	1984	105	K2
GStar A2	"	1985	103	K2
GStar A3	"	1989	136	K2
Galaxy I	Hughes	6/83	134	C
Galaxy II	"	9/83	74	C
Galaxy III	"	6/84	93.5	C
Galaxy IV	"	1989	140	C
Galaxy K1	"	1987	71	K2
Galaxy K2	"	1987	130	K2
Galaxy KSatI	"	1988	91	E
Galaxy KSatII	"	1989	93	E
Galaxy DBS 1,2	"	1989	101	K4

Satellite Name	Owner	Launch Date	Orbital Position (°W)	Frequency Band(s)
Marisat F1	Comsat General	1976	15	U,L,C2
Marisat F2	Comsat	1976	72.5	L,C2
Marisat F3	"	1976	176.5	U,L,C2
MMC1	Martin Marietta	1989	79	K2
MMC2	"	1989	125	K2
Orion I	Orion	1986	37.5	K
Orion II	"	1986	47	K
Orion III	"	1987	50	K
PanAmSat I	PanAM SC	1987	57	C,X2,K2
SBS I	SBS	11/80	99	K2
SBS II	"	9/81	97	K2
SBS III	"	11/82	95	K2
SBS IV	"	8/84	101	K2
SBS V	"	1986	122	K2
SBS VI	"	1987	62	K2
STC 1,2	Comsat	1986	110	K4
Satcom F1	RCA	12/75	—	C
Satcom F1-R	"	4/83	139	C
Satcom F2	"	11/76	—	C
Satcom F2-R	"	9/83	72	C
Satcom F3-R	"	11/76	131	C
Satcom F4	"	1/82	83	
Satcom F6	"	1986	67	C
Satcom F7	"	1989	62	C
Satcom K1	"	1985	85	K2
Satcom K2	"	1985	81	K2
Satcom K3	"	1987	67	K2
Spacenet I	GTE	5/84	120	C,K2
Spacenet II	"	1984	69	C,K2
Spacenet I	"	1985	87	C,K2
TDRS A	Spacecom	1983	41	C,K2,L
TDRS B	"	1985	61	C,K2,L
TDRS C	"	1984	171	C,K2
Telstar 1	AT&T	7/62	—	C
Telstar 2	"	5/63	—	C
Telstar 301	"	7/83	96	C
Telstar 302	"	1984	86	C
Telstar 303	"	1985	125	C

Satellite Name	Owner	Launch Date	Orbital Position (°W)	Frequency Band(s)
USA-Sat I	US-ISI	1988	56	K1,K2
USA-Say	"	1987	58	K1,K2
USSB 1	USSB	1988	110	K4
USSB 2	STC	1987	148	K2
Westar I	ASC/Western Union	4/74	—	C
Westar II	"	1974	—	C
Westar III	Western Union	8/79	91	C
Westar IV	"	2/82	99	C
Westar V	"	6/82	122.5	C
Westar VII	"	1990	144	C
Westar A	"	1988	73	K2
Westar B	"	1988	132	K2

Notes: 1. The frequency bands listed are as follows:

C—3.7 to 4.2 GHz
K1—10.7 to 11.7 GHz
K2—11.7 to 12.2 GHz
K4—12.2 to 12.7 GHz
E—20 to 60 GHz
U—0.3 to 1 GHz
L—1.0 to 1.17 GHz
X—6.5 to 10.7 GHz

2. ASC/WU is a joint venture of American Satellite Corporation (20%) and Western Union Telegraph Company (80%). ASC, in turn, is owned by Continental Telephone and Fairchild Industries.

3. These satellites are part of the Hughes HS 376 family which includes the SBS, Westar IV to VI, Anik C and D, Telstar 3 and Palapa B series.

4. Comstar D1 and Comstar D2, located closely together, operates as a composite satellite.

5. SBS is an abbreviation for Satellite Business Systems, a joint venture of IBM, Comsat and Aetna Life & Casualty.

6. Southern Pacific Communications, a private venture like ASC and SBS.

7. U.S. Satellite Systems, Inc.

Chapter

4

MESSAGES FROM THE SKY
The Earth Station

Antennas for receiving satellite broadcasts have become a prominent sign of our era. Large perforated, wire mesh and solid "dishes" are often seen at hotels and bars, at television stations, residential backyards as well as in the most remote rural areas. Although most people today realize that antennas are the "eyes" of satellite receiving stations, they are often perceived as mysterious and somewhat incomprehensible. Many questions arise. How many channels can an earth station receive? Why are these dishes so large and expensive? What is the difference between a square and round dish? Can a "do-it-yourselfer" build an inexpensive system? Will small 3-foot dishes that can be mounted in an attic soon be available? How does a satellite antenna have to be pointed to aim at the distant space vehicles? Will prices continue to fall? Is it all legal? Is it necessary to buy a special television or other equipment to see the satellite television programs?

The function of an antenna is to detect the minute signal received from a telecommunication satellite (Figure 4-1). Most dishes focus on one satellite at a time in order to concentrate these faint signals before funneling them into a device called a "feedhorn." The feedhorn then directs the microwaves into a small, box-like low noise amplifier, which further increases their power many thousands of times. Afterwards, these electrical messages are fed via cable lines to an indoor receiver/modulator which processes them into a form that can be understood by any television set. Microwave signals weaker than a tenth of a billionth of a billionth of a watt can carry vast amounts of information that can be detected and understood. Modern science has truly achieved wonders.

The Antenna

An antenna, often simply called a dish, must be of adequate size and quality to capture and sufficiently concentrate the faint signal

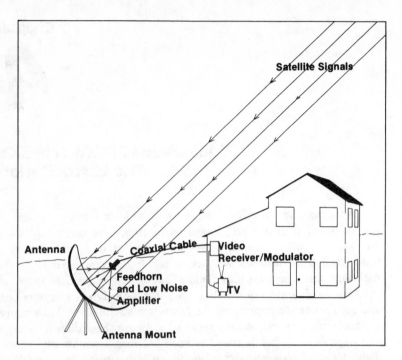

Figure 4-1. A Typical TVRO. An earth station includes an antenna and its mount, an actuator, a feedhorn, a low noise amplifier and a coaxial line to an in-doors satellite receiver and actuator controller.

from a distant satellite. It must also be accurately aimed toward these spacecraft so that other messages and noise from unwanted terrestrial and satellite sources are barely detected. The dish, about 40% to 50% of the cost of a complete earth station, must also be durable, aesthetically pleasing, reasonably priced and should maintain its accuracy for years.

Varieties of Antennas

Most microwave antennas used today in satellite earth receiving are built with combinations of circular or parabolic surfaces, which are geometrical shapes originally discovered by the ancient Greeks. Microwaves reflected from such surfaces will be concentrated to a point or to a series of points called "focal points." Even the original horn-shaped Telstar earth station, which weighed many tons and had to be rotated to follow its targeted satellite as it raced across the sky, had a parabolic cross-section.

Antennas can be grouped into two broad categories: single-beam and multiple-beam antennas. A single-beam antenna reflects signals from a satellite to a single feedhorn. In order to aim at a second satellite, the whole body of the antenna, with the feedhorn fixed in position, must be moved. Multiple-beam antennas use a reflective surface which focuses the signal to a series of points or a line where either one moveable feedhorn or multiple feedhorns are located. This permits simultaneous detection of more than one satellite.

The most familiar single-beam antenna, the prime-focus parabola, theoretically concentrates all incoming signals directed parallel to its axis to a single point. Incoming signals from off-axis directions when reflected will miss the focal point (see Figure 4-2 and 4-3). It is clear that the quality of a prime-focus as well as all other antennas is determined by how closely its surface approximates the designed shape.

In practice, a parabolic dish will not behave perfectly for two reasons. First, the equipment mounted at the antenna focus is spread around the focal point so that it will intercept some radiation from directions slightly off the target. Also, since the surface of an antenna is never a perfect parabola, surface imperfections cause errors in reflection so that some non-targeted signals are inadvertently detected and some targeted signals pass by unobserved. The quality of the prime-focus and all other antennas is determined by how closely the surface of a dish approximates its designed shape.

Note that a prime-focus antenna can be cut into any convenient shape. So if a round 15-foot prime-focus dish had its corners trimmed into a square measuring 9 feet on each side, it would still direct the intercepted radiation to a single point. The main difference between this dish before and after trimming would be in the surface areas. Thus, a 9-foot square antenna having an 81-square-foot cross-section would intercept slightly more signal than would a 10-foot round one having a 78.5 square foot cross-section.

A second type of antenna, the Cassegrain, also designed around a parabolic shape, reflects the signal a second time at the focal point to a location behind the dish (see Figure 4-2 and 4-4). The reflector at the focus is called a "parabolic subreflector." This type of antenna is often more expensive than an ordinary prime-focus dish, but can perform more effectively in certain situations. For example, in very hot climates the feedhorn and low noise amplifier operate more effectively being in a cool, shielded location behind the antenna.

There are also other less familiar designs of single-focus antennas. The horn reflector is shaped like a horn (Figure 4-5), and directs intercepted signals entering via a large aperture down the reflective body of the horn to a focus. The whole body of the horn performs the same function as the dish and feedhorn does for a prime-focus or

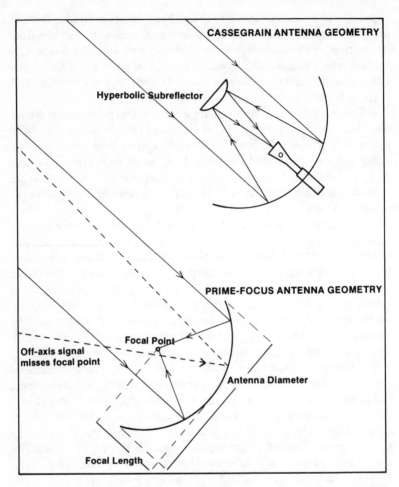

Figure 4-2. Prime-focus and Cassegrain Antenna Geometry. Signals originating along the main axis of either type of antenna are directed to the focal point. Off-axis signals miss the focus.

Cassegrain antenna. In some designs, the feedhorn and low noise amplifier are offset from the main axis. Both the Cassegrain and the prime-focus antenna can be used in such an offset-fed configuration. This setup has the advantage that the aperture of the dish is not blocked by the feedhorn and the low noise amplifier (see Figure 4-6 through 4-9).

Multiple-focus antennas allow more than one satellite to be detected from a dish fixed in one position. In contrast, a prime-focus dish has to be moved to receive signals from each satellite. Cable

Figure 4-3. A Prime-focus Antenna. This 10 foot dish is made from spun aluminum. *(Courtesy of Birdview Corporation).*

companies and other commercial earth station operators have the
option to own either two or more conventional antennas or one
multiple-focus antenna in order to simultaneously view transmissions from two or more antennas. Today, as the number of satellites
increases and as land becomes more and more expensive, a multiple-
focus dish can be advantageous.

Figure 4-4. A Cassegrain Antenna. This M/A COM 5-meter fiberglass dish has a subreflector mounted at the prime focal point. *(Courtesy of M/A COM, Inc.)*

The commonly encountered multiple-focus antennas incorporate variations of spherical and parabolic surfaces. In all cases, these dishes are aimed at the arc of satellites, and the reflected signals from each spacecraft are focused to a series of feedhorns. To

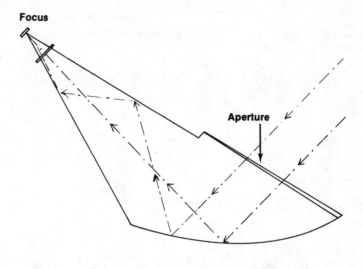

Conical Horn Antenna

Figure 4-5. Conical Horn Antenna Geometry. Signals are directed down the throat of the antenna to a low noise amplifier at the focus.

illustrate, the Simulsat antenna is a rectangular section of a sphere having its feedhorns mounted in a long box at the focal line. It can simultaneously detect up to twenty satellites within a 57 degree arc. In this case the feed is offset along the arc of satellites. The Torus antenna, which is a dual-curvature reflector, has a circular contour in facing the arc of satellites and a parabolic countour at right angles. It also focuses signals to a focal line and has an offset feed (see Figure 4-8).

How is the Quality of an Antenna Determined?

The quality of an antenna is simply determined by how accurately it concentrates microwaves intercepted from a targeted satellite to a signal point and by how well it ignores noise and unwanted signals coming from off-axis sources. Three interrelated concepts, antenna gain, beamwidth, and noise temperature are used to judge antenna performance. Two other descriptive measures, antenna efficiency and focal length to diameter ratio (f/D), also characterize system performance.

Antenna gain is a measure of how many thousands of times a satellite signal is concentrated by the time it reaches the feedhorn at its focus. For example, a typical well-built, 10-foot diameter, prime-focus dish will have a gain of 40 decibels or 10,000, which means

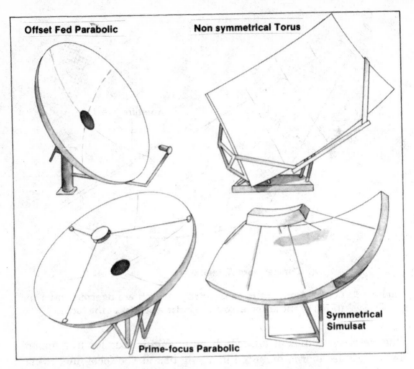

Offset Fed Parabolic

Non symmetrical Torus

Symmetrical Simulsat

Prime-focus Parabolic

Figure 4-6. Multiple-focus Antennas and a Prime-focus Antenna.

that the signal has been concentrated 10,000 times at the focal point.

Antenna gain depends upon three factors. First, gain increases as dish surface area increases because more satellite signal is intercepted. Thus, if the area of an antenna is doubled, the gain is doubled. Second, gain increases with frequency since higher-frequency microwaves can be more easily focused into straight lines like beams of light so that more signal is captured. The simple rule is that antenna gain is proportional to the square of the microwave frequency. For example, a signal with twice the frequency will be captured by an antenna with four times gain. To illustrate, if the gain is 10,000 when a signal of 5 gigahertz is received, then it will be 40,000 at 10 gigahertz). Third, gain is determined by how accurately the surface of an antenna is machined to exactly a parabolic or other selected shape. Even small surface distortions can cause an antenna to loose substantial amounts of gain (see Table 4-1). Therefore, a dish that has ripples in its surface will behave more poorly than one that is smooth and is closer to its designed shape.

Figure 4-7. Multiple-focus Antenna. Two feedhorns located on the focal line each detect signals from different geosynchronous satellites.

The beam pattern of an antenna measures how well it can "see" into a very narrow region of space and ignore off-axis signals and noise. It is crucial that dishes have such very focused vision because satellite separated by 4 degrees or less and located more than 22,300 miles away appear to be very close together. Signals from a chosen satellite are detected by the the "main lobe," while extraneous off-axis signals are seen by "side lobes."

TABLE 4-1. GAIN REDUCTION VS. SURFACE DISTORTION	
Distortion (inches)	Gain Reduction (%)
0.01	0.2
0.05	1.7
0.10	4.4.f
0.25	28.8

Each antenna has a characteristic beam pattern Antenna beamwidth is then defined as the width, in degrees, between points on either side of the main lobe where power is received at half of that

Figure 4-8. The Torus Multiple-focus Antenna. This antenna has a circular contour facing along the arc of satellites and a parabolic contour at right angle. As a result it directs signals to a focal line. *(Courtesy of SatCom Technologies, an RSI Company)*.

detected along the main axis. Antenna beamwidth decreases as both the frequency of the satellite message and the antenna diameter increase. A poor-quality dish will have a relatively wide main lobe and larger side lobes. As a result, it will see relatively more noise

Figure 4-9. The Simulsat Multiple-focus Antenna. The Simulsat antenna is a rectangular section of a spherical surface. *(Courtesy of Antenna Technology Corporation).*

and less on-axis signal. Note that when less reflector aperture is blocked, as in the case of antennas with offset feeds, side lobes can be lower.

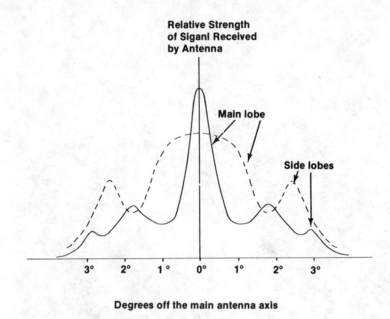

Figure 4-10. Comparison of Antenna Lobe Patterns. The better antenna has a more narrow main lobe and smaller side lobes.

Beam patterns always deviate from those that are theoretically predicted. Side lobes increase with reflector surface irregularities. Similarly, reflections from the feedhorn/low noise amplifier structure, the edges of the dish, and the surrounding terrain increase side lobes and impair dish performance. In general, sides lobes can be minimized if a dish has its feedhorn located relatively closer to the antenna surface so that it sees less of the surrounding environment (see Figure 4-11). A similar result can be obtained if a non-reflective shroud is affixed around the rim of an antenna which is similar to putting blinders on a horse so it will see only in one direction, straight ahead (see Figure 4-13).

The difference between an excellent and a poor-quality antenna is easily seen in their beam patterns. It is important to understand that even apparently small changes in design can effect the "vision" of a dish. For example, square and round prime-focus antennas having identical surface areas and gain, can have quite different lobe patterns.

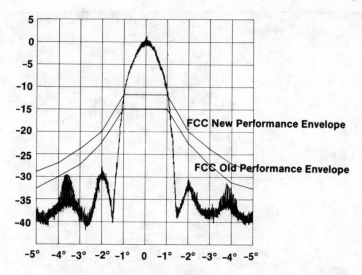

Figure 4-11. FCC Antenna Performance Envelope and Actual Performance Chart. The side lobes of a quality antenna should fall below those recommended by the FCC.

Antenna performance is also characterized by the unwanted noise that is detected along with the targeted satellite signal. Off-axis signals and noise coming from many man-made or natural sources is unavoidably detected by antenna side lobes. Natural sources of microwave noise are usually those which affect reception of satellite broadcasts since the frequency of noise coming from man-made devices (e.g., electrical motors, fluorescent lights) is too low to affect satellite transmissions. It is clear that if an antenna has a narrow beamwidth and acceptable small side lobes it will detect relatively small amounts of noise.

The overall performance of an antenna and its feedhorn/LNA is also characterized by a quantity called antenna efficiency. This is a measure of what percentage of the incoming radiation is actually captured by the dish system. Typical antenna efficiencies vary from 55% to 65%. In other words, 35% to 45% of the incoming signal is lost before being converted to an electrical signal in the low noise amplifier.

A quantity called focal length to diameter ratio, f/D, is used to decribe how closely the feedhorn is located to the antenna surface. Typical f/Ds range from approximately 0.27 to 0.41. A deep dish, meaning one which has the feedhorn located more closely to dish

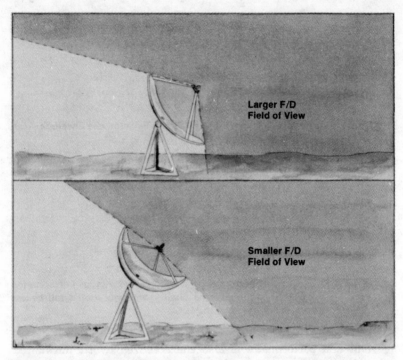

Figure 4-12. The Effect of Focal Length to Antenna Diameter Ratio. An antenna having a lower f/D sees less of the "noisy" surrounding terrain because the low noise amplifier is located more closely to the reflective surface.

surface, has f/D less than approximately 0.35. Deep dishes can have lower side lobes since the feedhorn is shielded from the surrounding terrain by the antenna edges.

What is Interference?

Interference occurs when unwanted signals are received along with the desired satellite message. In some cases, the garbling of the message can be so severe as to completely ruin a broadcast. Four types of interference can be identified.

The most common and irritating form of interference, unaffectionately known as terrestrial interference or TI, is caused when microwave signals using the same or an adjacent frequency band are picked up by the dish and earth station electronics. When these interfering signals are sufficiently powerful they are detected by a TVRO antenna and can ruin a broadcast. Surprisingly, in some situations, moving an antenna just a few feet can significantly reduce interfering signal levels. Of course, in those situations where interference is very powerful, a TVRO may have to be installed in a specially protected area, often at high cost.

The FCC frequency allocations in the microwave region near those of satellite transmissions identifies some possible sources of interference (see Table 4-2). Line-of-sight relays used by common carriers transmitting telephone, data and video information that share common frequencies are the prime contributors of interference. This is so even though the carrier waves transmitted from satellites are specially altered to give them a different identity from such transmissions.

In some situations, stray signals from adjacent satellites or from uplinks targeting adjacent satellites can be detected by a TVRO. This troublesome interference can impair reception of the desired broadcast. Two types of adjacent satellite circuit interference can occur, uplink and downlink interference (see Figure 4-13). The downlink variety occurs when a poor-quality dish having relatively large side lobes or an undersized dish having a wide beamwidth detects unwanted signals from an adjacent satellite. Uplink interference occurs when stray power from an uplink antenna pointed towards an adjacent satellite is detected by the target satellite and subsequently relayed to the TVRO antenna. In all cases, interference will come from transmissions occupying the same bandwidth as the one being viewed. Thus, if transponder 10 messages are being viewed, only signals on transponder 10 from an adjacent satellite or uplinked to an adjacent satellite would be powerful enough to potentially cause problems.

A third possible source of interference is from channels adjacent to the selected transponder. However, this would only occur with the use of a low-quality feedhorn that discriminates poorly between signals of opposite polarization. For instance, in a well-designed TVRO station, channel 12 should not receive "crosstalk" from channel 11 or 13.

Another predictable form of interference is caused by the sun. Twice a year the sun lines up directly behind each satellite for periods of approximately ten minutes per day for two or three days. Since the sun is a source of massive amounts of microwave noise, no transmissions can be received from satellites during these "sun-outage" times. This unavoidable type of interference should be expected during the normal course of operation of an earth station.

The FCC has issued two rules aimed at reducing problems relating to interference. However, since the 1979 deregulation of TVRO earth stations, owners and manufacturers of antennas are not man-

TABLE 4-2. MICROWAVE FREQUENCY ASSIGNMENTS AS POSSIBLE SOURCES OF TERRESTRIAL INTERFERENCE

Frequency Band	Potential Source of Microwave Interference
0.960—1.350	Land-based air navigation systems
1.350—1.400	Armed forces
1.400—1.427	Radio astronomy
1.427—1.435	Land-mobiles: police,fire,forestry,railway
1.429—1.435	Armed forces
1.435—1.535	Telemetry
1.535—1.543	Satellite—mobile maritime
1.605—1.800	Radio location
1.660—1.670	Radio astronomy
1.670—1.700	Meteorological radiosound
1.700—1.710	Space research
1.710—1.850	Armed forces
1.990—2.110	TV pickup
2.110—2.180	Public common carrier
2.130—2.150	Fixed point-to-point (private)
2.150—2.180	Fixed—omnidirectional
2.180—2.200	Fixed point-to-point (public)
2.200—2.290	Armed forces
2.290—2.300	Space research
2.450—2.500	Radio location
2.500—2.535	Fixed satellite
2.500—2.690	Fixed point-to-point (private)
2.655—2.690	Fixed satellite
2.690—2.700	Radio astronomy
2.700—2.900	Armed forces
2.900—3.100	Maritime radio navigation
3.300—3.500	Amateur radio
3.700—4.200	Common carrier (telephone)
	Earth Stations
4.200—4.400	Altimeters
4.400—4.990	Armed forces
4.990—5.000	Meteorological & radio astronomy
5.250—5.650	Radio location (coastal radar)
5.460—5.470	General radio navigation
5.600—5.650	Maritime radio navigation
5.600—5.650	Meteorological—ground based radar
5.650—5.925	Amateur
5.800	Industrial and scientific equipment
5.925—6.425	Common carrier and fixed satellite
6.425—6.525	Common carrier
6.525—6.575	Operational land and fixed satellite
6.575—6.875	Non-public point-to-point carrier
6.625—6.875	Fixed satellite
6.875—7.125	TV pickup
7.125—8.400	Armed forces
8.800	Airborne Doppler radar

Figure 4-13. Cassegrain Antenna with Protective Skirt. This non-reflective rim shields the dish from detecting unwanted interference and noise. *(Courtesy of Scientific Atlanta).*

dated to follow these regulations. Nevertheless, they are based on careful study and are very useful guidelines. First, a set of well-defined standards restricts the amount of allowable side lobes that an antenna can exhibit (Figure 4-10). Second, a procedure known as a "frequency coordination" has been established to protect broadcasters and other communicators who operate commercial systems.

Frequency coordination, a three-part process, begins with an on-site engineering analysis to measure levels of microwave interference. Next, a coordination notice is issued to inform other local microwave users of the intent to install the facility. Third, the FCC issues an operating license so future microwave users can easily check to see where other users are located.

How Large Must an Antenna Be?

Perhaps the most regularly asked question is why such a large antenna must be used. Today, small 3-foot dishes that are capable of receiving higher-frequency television broadcasts directly from satellites are often compared to larger, more common antennas. To answer this question, we can analyze the five factors influencing the antenna size requirements: satellite footprint, detected noise, antenna diameter and gain, antenna beamwidth and interference. These all determine a minimum ratio of signal-to-noise required so that earth station circuits can deliver a clear television picture.

Antenna size is directly related to the intensity of the satellite signal at the TVRO site. Doubling this power simply means that antenna surface area can be cut in half. This would be equivalent, for example, to replacing a 10-foot dish with a 7-footer.

The amount of power received at any point on earth is defined by the footprint of a telecommunication satellite. Signal power could be increased up to a point which is limited by the capacity of the launch vehicles and by the capability of the on-board solar cells. Certainly satellite power levels are inching upwards as heavier, more complex spacecraft are being launched into orbit each year. In general, newer satellites do have more available power. However, since the average lifetime of satellites is about eight years, it will be many years before required antenna size will decrease substantially due to increases in EIRP. Another alternative for increasing satellite EIRP is to use more directional zone or spot beams to concentrate the available power. This allows use of smaller TVRO dishes given that a fixed amount of power is available. However, using such localized footprints which target a limited geographical area is acceptable only for certain types of broadcasts.

Antenna beamwidth is also an important and limiting parameter in selecting dish diameter. Since beamwidth decreases as diameter increases and since C-band satellites are being spaced at 2 degree intervals, a dish having a diameter less than approximately 6 feet will detect two or more closely spaced vehicles. (This point is discussed in more detail in Chapter 7). Much smaller antennas can be used for higher frequency Ku-band broadcasts because power levels are higher and beamwidth is much more narrow.

Required dish size is also determined by how much noise is detected by the antenna/feed/LNA system. The higher the level of noise "seen" by a dish, the higher antenna gain must be in order to deliver an adequate electronic signal. Better quality dishes having smaller side lobes and a more narrow main lobe intercept less noise. Note that as antenna diameter is increased, although both gain and detected noise increase, the latter increases at a slower rate. Also, in higher latitudes, an antenna must be aimed at a lower angle, the "look angle," to point towards the arc of satellites. It therefore sees more of the warm, "noisy" ground as well as a thicker layer of warm atmosphere. All of the improvements in performance that increase gain and lower antenna noise allow use of smaller antennas, and vice versa. Similarly, a location that has unnecessary sources of interference will need either a larger or a higher-quality antenna to maintain an adequate signal-to-noise ratio.

All these factors taken together determine the minimum dish size that can deliver adequate performance. For conventional 6/4 gigahertz broadcasts, a minimum antenna diameter of from 6 to 15 feet is required in the continental United States. Higher-frequency broadcasts are most often relayed to more restricted areas and therefore have higher EIRPs. Therefore, much smaller antennas can be used to detect the so-called direct broadcasts that have frequencies in the range of 12 gigahertz.

Of course, it is not safe engineering practice to use the smallest possible antenna with no margin for error. In time, even the best antenna will age and its performance will deteriorate. In addition, satellite transponders tend to degrade over their lifetime. This causes EIRP to slowly decrease. Note that the inevitable small movements in the orbital position of satellites or slight misalignments in dish aiming and feedhorn position can reduce gain and lower picture quality.

Antenna Construction and Installation

The materials, construction, and installation must be carefully considered if antennas are to function properly over a long lifetime. Microwave dishes made of fiberglass, aluminum, steel, and even of wood or metal screen may perform well when new. However, the performance of dishes which have been poorly manufactured, designed or constructed from less durable materials may quickly degrade after they are exposed to the elements for sufficient time. For example, wood may warp, or a thin screen mesh can be easily damaged by rocks and hail. A thin-wall dish with poor structural support will also be more likely to warp. Allowances must be made for the expansion and contraction of aluminum and steel antennas so that their accurate shape will be retained. Note that high-quality, well-designed dishes may be constructed from one or many pieces.

An effective antenna surface is not necessarily guaranteed by the use of a quality material and design, but also depends upon an excellent manufacturing process. All antennas must be capable of reflecting microwaves. Dishes of wire mesh, aluminum, and steel are metallic and thus already reflective to microwaves, but fiberglass dishes must have metal-based coatings incorporated into their design. The challenge of manufacturing a fiberglass antenna is to embed wire screen, aluminized mats, or other metallic materials in the fiberglass resin while retaining an accurate surface. The details of this manufacturing process determines the lifetime and durability of the final product.

The overall performance of an antenna varies with how well the installation has been planned and implemented. For example, shadowing of a satellite signal by local obstructions such as trees or buildings can degrade performance as much as poor alignment can. On the other hand, an antenna mounted on the ground with foliage and buildings blocking out all but the desired satellite signal may dramatically eliminate interception of interference and ground noise and may therefore improve overall system performance. This strategy is sometimes used when antennas are installed in pits where the view is restricted to only the arc of satellites. Unlike a regular TV antenna that must be mounted very high to avoid obstructions and receive a clear signal, a satellite receiving antenna needs only a clear "view" of the satellites, so dishes can generally be mounted at ground level where winds are usually lower than on roofs. A closed attic would also be a convenient and protected installation site. Unfortunately, this is not possible because most of the signals would be absorbed by the roof materials. Even using a glass cover to protect the surface from rain or wet snow is not a reasonable alternative, since glass substantially reduces the power of 4-gigahertz satellite signals.

Are Earth Station Antennas Safe?

Earth receiving stations are perfectly safe. The levels of microwave radiation present even after amplification by the antenna and the low noise amplifier at its focus are negligible. Antennas do not get hot, do not fry birds, and do not transmit dangerous radiation. The voltage delivered to power the amplifier at the focus is at the highest only 24 volts, about twice that of a car battery. There is no danger, except perhaps from a poorly mounted antenna falling over in an extremely strong wind.

Antenna Mounts

Earth station antennas must be securely mounted and capable of being aimed accurately towards any chosen telecommunication satellite. Using a mount that provides stability and pointing accuracy is a critical part of a quality TVRO, because well-designed dishes with narrow beamwidths target very small portions of the sky. For example, a 5-meter antenna with a beamwidth of one degree sees only twenty-six ten thousandths (1 in 26,000) of the visible sky. Therefore the satellite signal would be lost if this dish moved even a relatively small distance off target. Without an adequately stable mount, such movement is possible when winds blow across the large sail-like, surface areas of the antenna.

Antenna supports must be strong, rigid structures, firmly attached to both the dish and the installation location. They can be bolted onto concrete pads, mounted on well-planted poles or affixed onto carefully designed roof structures. In general, an antenna should be secured as close to the ground as possible, preferably in a protected area. Dishes mounted on rooftops must be attached to a structurally sound building able to resist strong uplifting winds. Of course, the smaller the antenna, the easier it is to mount on a roof or any other chosen location.

Types of Mounts

There are two broad classes of structures used for mounting earth station antennas. One consists of those assemblies having two axes of movement at right angles to each other, such as azimuth-over-elevation (az-el) or x-y mounts. The other class of structures, polar and modified polar mounts, have just one axis of movement used to track the geosynchronous arc of satellites.

X-y and az-el mounts are similar except for the placement of their two axes (Figure 4-14 and 4-15). The x-y has one axis parallel to the ground like the parallel bars of a gymnast. The az-el has its lower axis perpendicular to the ground so it has movement like a spinning figure skater. In the United States, the elevation angle adjustment required for both these types of dual axis mounts ranges from 7 degrees in northern Maine to 25 degrees in southern Florida.

Installation of such mounts is rather easy because the only adjustment necessary before securing the assembly is to align the

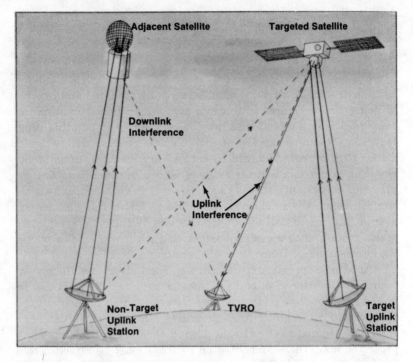

Figure 4-14. Adjacent Satellite Interference. Stray signals from either uplink or downlink sources can cause interference.

body of the mount in a level plane, with the midpoint of either the azimuth or y-axis pointing to near the center of the satellite arc. Although these mounts are simpler to manufacture and install than are polar mounts, they are generally more cumbersome to use in scanning the arc because two axes of movement must be carefully adjusted for each satellite. Recently, easy-to-use-and-install computer controlled two-axis mounts for positioning even the smallest dishes have been introduced.

Polar mounts have the advantage of being able to track satellites from the eastern to the western sky by movement along only one axis, the polar axis (see Figure 4-16). The development of these single-axis mounts was motivated by efforts to reduce the costs and increase the flexibility of earth stations.

Installing polar mounts requires care. The body of such a mount must first be carefully aligned towards true south. Then the polar axis angle is adjusted equal to the site latitude so the polar axis lies parallel to the earth's true north/south axis. Finally the declination angle set so that the antenna is tilted slightly downwards

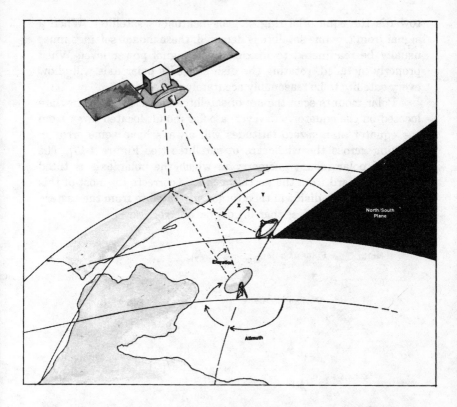

Figure 4-15. Geometry of Az-El and X-Y Mounts. Two axes of rotation are needed in both these devices to target a communication satellite.

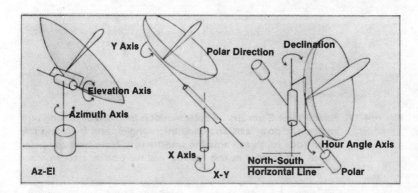

Figure 4-16. Schematics of Antenna Mount Geometries. This diagram shows how the axis or axes are rotated for these three classes of mounts.

towards the equatorial ring of geosynchronous satellites. When a signal from any one satellite is detected, these mount settings must usually be readjusted to maximize the signal power level. When properly installed, rotating the dish along its polar axis will allow every satellite to be reasonably accurately targeted.

Polar mounts scan the arc of satellites perfectly when they are located on the equator. However, a polar mount located away from the equator at non-zero latitudes will always have some error in tracking across the whole arc of satellites (see Figure 4-17). The modified polar mount geometry in which the polar axis is tilted slightly forward from the north/south axis corrects for most of this tracking error. A dish will then target any satellite from the eastern to western sky with less than 0.1 degree of error.

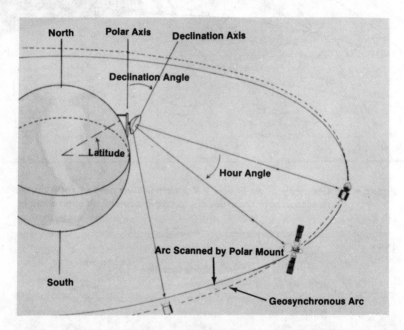

Figure 4-17. Polar Mount Geometry. A polar mount is installed by setting two angles, the polar axis and declination angles, and by facing the mount due south towards the satellite belt. Note that any polar mount not located at the equator will have some error in scanning the arc.

Antenna Actuators

Today most home satellite systems are equipped with actuators which automatically aim a dish towards any chosen satellite. These are either linear actuator which are motor driven jacks having one end attached to the mount and the other to the dish, or horizon-to-horizon mounts consisting of a gear assembly located just behind the mid-point of the antenna structure.

Commercial "headends" located, for example, at condominium complexes or on the premises of cable TV companies often use larger antennas which are fixed towards just one satellite. Occasionally, such organizations will use fixed, multi-focus antennas which simultaneously target two or more spacecraft.

The Feedhorn and its Support Structure

Signals concentrated by an antenna are captured by a device called a feedhorn which then channels the microwaves into a low noise amplifier. Just as the quality of an antenna determines how well signals are "seen" and defines its beam pattern, so does a feedhorn "illuminate" a dish. A theoretically perfect feedhorn illuminates only the dish surface so that it sees reflected signals coming from a direction along the antenna axis but ignores signals coming from anywhere else. In reality, a feedhorn unavoidably detects noise and signals from the surrounding terrain and off-axis directions as well as the desired signal.

Feedhorn and antenna designs must complement each other in order to capture as much signal and as little noise as possible. Using a poor-quality feedhorn can degrade the performance of an otherwise excellent dish. A low-quality or an improperly located feed can also dramatically increase side lobes.

How Does a Feedhorn Work?

Feedhorns are based on waveguide designs which are metal, hollow pipes of circular, rectangular, or other cross-sectional shapes that transmit microwaves (see Figure 4-17). Waveguides are analogous to fiber-optic cables that transmit light along their lengths. Microwave signals can be transmitted through a waveguide only when the wavelength is shorter than one half of the dimensions of the waveguide. This very useful property allows an earth station to discriminate signals of vertical and horizontal polarization. For example, if microwaves having a wavelength of 1 centimeter reach a waveguide having a cross-section of 2 by 0.3 centimeters, only those waves polarized in the wider dimension, 2 centimeters, will pass through.

All feedhorns channel microwaves reaching the focus of an antenna into a waveguide that accepts only one of the two polarizations. In more primitive earth station designs a single rectangular waveguide was used and the whole feedhorn/low noise amplifier assembly was rotated to accept either horizontally or vertically polarized signals. Modern feedhorns can select between horizontally and vertically polarized signals in a much more elegant fashion. Usually a DC motor or a servomotor rotates a small metal probe located in the feedhorn throat. Other less common designs use non-moving, solid-state diodes or ferrite devices controlled by electrical current to select between signal polarities.

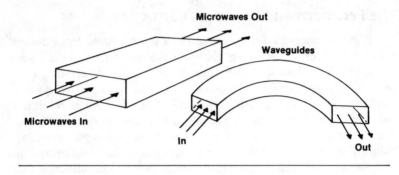

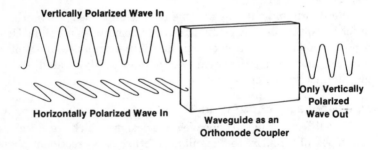

Figure 4-18. Waveguides. Horizontally and vertically polarized signals can be separated by waveguides of the appropriate dimensions.

Some feedhorns incorporate a waveguide design called an "orthomode coupler," which uses two rectangular boxes at right angles to separate horizontally and vertically polarized signals for relay to two separate low noise amplifiers. Such a dual feedhorn/low noise amplifier assembly can simultaneously receive all channels from any one satellite at once without the need for mechanical or solid state polarity selection equipment (Figure 4-19).

The choice between a dual- or single-polarization feedhorn is based both on convenience and on the cost difference between a more expensive feedhorn and an extra low noise amplifier. Nearly all

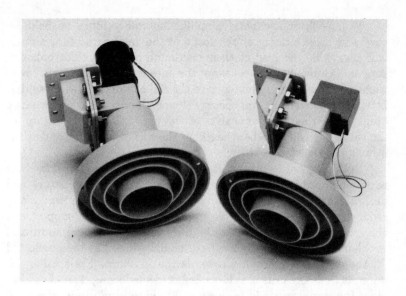

Figure 4-19. Feedhorns. Probes Inside the Chaparral Polarotor™ I and II are driven by a servo motor and a DC motor, respectively. Both of these polarity selection devices rotate to their final position in less than half a second. *(Courtesy of Chaparral™ Communications).*

home satellite systems use a feedhorn with selectable polarity. The feedhorn probe must be carefully aligned with the direction of microwave polarization so that signal power is maximized and so that signals of the opposite polarization are rejected. Most satellite receivers have a skew adjustment knob which allows for fine tuning of the feedhorn probe. This is necessary because the directions of signal polarization vary somewhat from satellite to satellite.

What are the Types of Feedhorns?

The most commonly encountered feedhorn is a "scalar" design. This type has a series of concentric rings that direct signals from an antenna into a central waveguide. More primitive feedhorns used a "cow-bell" device which was shaped like an inverted bell and which directed radiation into a smaller waveguide at its base. The feedhorn used with a Cassegrain antenna is a specially designed waveguide which directs signals from the subreflector at the focus to a low noise amplifier sometimes located behind this dish. Buttonhook-type feedhorns are attached to the dish center and direct radiation down a long curved waveguide to an LNA in a rear location.

The feedhorn/low noise amplifier assembly must be held securely in place exactly at the focus of the antenna so that signals will always be received at their maximum power. A "buttonhook" support is affixed to the center of the antenna. Such supports sometimes must be securely fastened to the rim of a dish by wires so that little or no movement will occur. A "moon-lander" (also known as a "milk stool") assembly consists of three or four poles that are attached at one end to either the rim or the surface of a dish and at the other end to the feedhorn.

Both types of feed supports must be carefully designed in order to be compatible with the overall antenna/feedhorn structure and design. Any type of support can block the incoming microwaves and reduce the amount of available signal. The buttonhook support and the moon-lander attached on the dish surface have the additional problem of shadowing some of the signal which has been reflected off the dish and is heading towards the feedhorn. The smaller the antenna the more crucial is the amount of shadowing. It is clear that since the signal received from a satellite is so very low in power, all the components must be designed to work as a well-matched team.

Figure 4-20. Dual Polarity Feeds. Both vertically and horizontally polarized signals can be simultaneously detected using these feeds which have two output ports. (Courtesy of Chaparral™ Communications).

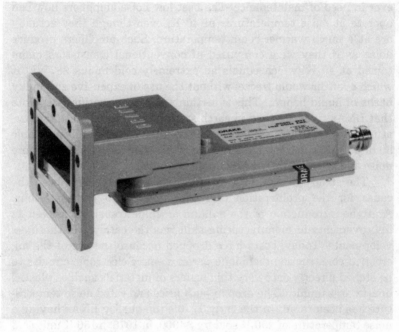

Figure 4-21. A Low Noise Amplifier. This 85 K low noise amplifier is the first active electronic component of a satellite receiving station. (Courtesy of R. L. Drake Co.).

The Low Noise Amplifier

The low noise amplifier (LNA) receives a concentrated signal from the antenna/feedhorn. It has the crucial function of further increasing signal strength by a factor of approximately 100,000 while maintaining the noise level very low so that a sufficiently powerful message will reach the satellite receiver. The antenna, feedhorn and low noise amplifier working in unison are the most important parts in determining how well an earth station works. The LNA is the first "active" or electronic component in the processing of a satellite signal.

Improvements in LNA design have been made possible by recent advances in transistor technology. More primitive LNAs were very bulky and often had to be immersed in baths of extremely cold liquid nitrogen or helium to keep noise levels acceptably low. The development of the gallium arsenide field effect transistor, a special low noise transistor, has made the modern LNA possible. When properly installed, the LNA has a very long lifetime and is hardly ever in need of maintenance. The best low noise amplifiers now can operate at noise temperatures of 40 K, even though they actually are at a much warmer room temperature. Such amplifiers produce noise as if they were composed of conventional transistors maintained at 40 K (which equals an extremely cold minus 387° F at which even air would freeze) without the use of expensive and bulky baths of liquid helium! This is certainly a great scientific feat. Note that LNAs typically used in earth stations with rated noise temperatures of 80 to 100 K are less expensive than commercially available 50 or 60 K low noise amplifiers because they introduce slightly more noise and are of slightly lower quality.

The steady decline in low noise amplifier costs have been the cause for the proliferation of moderately priced earth stations. And the introduction of the gallium arsenide transistor as well as improvements in manufacturing skill was the catalyst in these developments. Today, costs have dropped because the use of the microstrip construction technique permits entire electronic circuits to be etched directly onto very thin wafers of materials such as plastic, quartz, or sapphire. The drop in both price and rated noise temperature has been rapid. In the early 1970's top-quality LNAs having a noise temperature of 250 K sold for $4000; in 1976, a 150 K unit sold for $4000. At the end of 1977, 100 K LNAs were selling for less than $4000, and 150 K units for below $2000. By 1982 the consumer could purchase an LNA rated at 100 K for below $500. Today it is not uncommon to find 60 or 70 K LNAs selling for less than $200!

How Does an LNA Work?

LNAs are designed to uniformly amplify signals over the whole C-band range while introducing a minimal amount of extraneous noise. The LNA is analogous to a stereo pre-amplifier in that it provides the first critical electronic amplification. The LNA is bolted directly onto the feedhorn so that a minimal amount of incoming signal will be reflected and lost. Inside the small metal box that houses the LNA, microwave signals are usually amplified in three or four stages to achieve the overall 100,000 times necessary increase in signal strength.

A cable attached to the rear of the LNA transmits the message to the next stage, the downconverter, which is mounted at the dish and which is a component of the satellite receiver. A second cable then relays the signal to the indoors electronics. These cables also transfer approximately 15 to 24 volts DC from the satellite receiver to provide power for amplification and downconversion. Amazingly, the output signal leaving the LNA is still very weak at a power level of approximately one hundredth of a millionth of a watt!

Until the 1980s most low noise amplifiers simply amplified the signal received from the satellite and relayed the whole 3.7 to 4.2 gigahertz band of frequencies (having a bandwidth of 500 megahertz) to the satellite receiver. These satellite receivers had built-in downconverters. Modern TVROs "downconvert" as well as amplify the signal at the dish. Downconversion, which means lowering the frequency of the carrier wave, is an important step because very high microwave frequencies can be processed only with extremely expensive and sophisticated electronics. The original audio and video information can be more easily and cost effectively recovered at conventional intermediate frequencies (IF).

Today, some LNAs combine the functions of amplification and downconversion into one component. A "low noise converter" (LNC) downconverts each channel in the 3.7 to 4.2 gigahertz signal to an IF frequency range. For example, one model lowers the frequency of each channel in turn to the 50 to 90 megahertz band. Another type of low noise amplifier, called a low noise block converter (LNB), simultaneously lowers the microwave carrier frequency of the whole 500 MHz wide block of satellite channels to the range in which signal demodulation is performed (see Figure 4-22 and 4-23).

Low noise amplifiers have one weakness in their overall durability. Their special transistors act like fuses and burn up if the voltage powering their performance exceeds operational limits. Half of all LNA failures result from excessive voltages from unpreventable lightning strikes before the normal lifetime caused by wear and tear is reached.

How Does Heat Effect LNAs?

How well a low noise amplifier performs is determined to some degree by the temperature of the environment in which it operates. If the temperature rises from 77° F to 95° F, the gain of a 75 K LNA could decrease by almost 3% and the noise temperature could increase by just under one degree Kelvin. Using a Cassegrain an-

Figure 4-22. LNA and LNB. The LNB on the left downconverts the whole satellite signal frequency band as well as providing amplification. A
standard LNA is shown on the right for comparison. *(Courtesy of
M/A COM, Inc.).*

tenna in a hot desert climate is advantageous, since the LNA is
mounted behind the dish out of the direct glare of the sun and therefore remains cool. For example, if a 120 K LNA were exposed to
direct sunlight and heated from its normal 68° F operating temperature to 180° F, it would behave like a 125 K unit by adding more
noise to the signal.

LNAs can be protected from the heat in a number of ways. A
cover can be used to house the electronics of the LNA to help keep
temperatures low. These containers are often hermetically sealed
by rubber or enamel paint to shield the sensitive circuits from
adverse climatic conditions. There is an additional clever method for
improving the performance of an LNA. The first sensitive stages of
the amplifier can be "spot-cooled" by a thermoelectric device operating at a low power of 2 to 5 watts and placed on the surface the LNA

Figure 4-23. Low Noise Converter. This LNA amplifies and downconverts one channel at a time from the satellite broadcast. *(Courtesy of Avantek Corporation).*

metal box. For example, if a 120 K LNA were cooled to 0° F, its effective noise temperature would drop to 117 K.

How is the LNA Selected?

The choice of an LNA is determined by the cost and design tradeoffs between various system components. The characteristics of the antenna, feedhorn and LNA determine how much signal is captured from any satellite. However, if the signal-to-noise ratio received from the antenna and feed is too low, the highest quality low noise amplifier cannot rectify the problem, and picture quality will be poor. In most cases, an 80 to 100 K LNA can contribute enough amplification with low enough noise so that the message sent via cable to the receiver is of adequate quality.

In general, the lower the LNA noise temperature, the higher will be its cost. Most earth stations now use devices rated from 60 K to 100 K. However, other factors enter into the retail price. An LNA that has a higher gain or has surge protection to isolate its sensitive components from power surges will usually cost more. The method of rating LNAs can also effect price. For example, some brands rated at 90 K are guaranteed to have noise temperatures below this figure, often by as much as 15 or 20 K; other less expensive LNAs rated at 90 K are usually very close to or even slightly higher than this value in their noise temperatures.

From the LNA to the Receiver—Coaxial Cable

The antenna/feedhorn/LNA and downconverter are connected to the indoors receiver and its built-in modulator by a special type of wire called coaxial cable. Single wires of copper or aluminum ate adequate for conducting electricity at the lower frequencies encountered in most familiar electrical devices. However, when higher-frequency electrical signals are relayed on wires, these single strands of metal behave like antennas by radiating away most of the power. With exceptionally high frequency microwaves, like those employed in satellite communication, specially designed coaxial cables must be used to prevent almost complete loss or attenuation of the signal enroute to the video receiver.

The basic design of coaxial cables is standard. They are constructed having a piece of wire surrounded by an insulator, which itself is encircled in a wire sheath that is connected to ground (see Figure 4-24 and 4-25). This grounded layer of metal inhibits the ability of the central conductor to act as an antenna. Therefore high-frequency messages are lost at a much lower rate as they travel down the coaxial cable. The central conductor is also well shielded from unwanted external signals.

Figure 4-24. Coaxial Cable. The tubes surrounding the central conductor in this "coax" are filled with non-conducting air. (Courtesy of M/A COM, Comm/Scope Marketing, Inc.).

The cost of coaxial cable generally increases as its upper frequency limit increases. This is so because it must be designed to more exacting specifications and because higher quality sheathing necessary to prevent wires carrying high-frequency signals radiating away large amounts of power is more expensive. Also any error made when intalling TVRO equipment takes a higher toll and causes more deterioration in performance if higher-frequency signals are used. To illustrate, if "coax" were not properly connected to the rear of an LNA there would be over 70% more losses at 12 GHz that at 4 GHz. It is clear that one reason for downconverting to lower frequencies at the dish is to minimize problems caused when work-

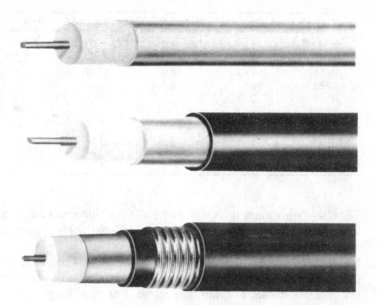

Figure 4-25. Coaxial Cable. These three examples of coaxial cable for high frequency signal conduction are protected to increasing degrees (top to bottom). The middle has a polyethylene jacket. The lower has two jackets as well as a corrugated chrome plated steel armor to prevent the cable from being excessively bent. *(Courtesy of M/A COM, Comm/Scope Marketing, Inc.).*

ing with high microwave frequencies. Using lower frequency coax has the added advantage of reducing the amount of money spent on long coax runs. For example, RG-214 coax used to carry the C-band signals from the LNA to the downconverter is more than ten times the cost of RG-59. This less expensive coax can effectively transmit the lower frequency signal from the downconverter over distances which are sometimes hundreds of feet from the dish site to the satellite receiver (see Table 4-3).

Coaxial cable is usually classified as either hardline coax, or foam or air dielectric coax depending upon the construction of its dielectric sheathing material. Coax has one or two pliable metal grounded layers wrapped around a plastic dielectric. Foam or air dielectric coax uses either foam or compressed air as the dielectric material and is generally lower loss and more expensive than ordinary coax. Hardline is similar in construction to either of the previous two types except it has a more rigid, even lower loss metal sheath. Hardline is usually employed in the main truck lines of cable or other communication systems.

TABLE 4-3. CHARACTERISTICS OF COMMONLY USED COAX

| Cable Type | Signal Loss (dB/100 feet) | | | Imdepance (ohms) |
	100 MHz	1450 MHz	4 GHz	
RG-59	3.40	11.0	N/A	75
RG-6A	2.70	8.7	N/A	75
RG-11	2.30	7.0	N/A	75
RG-8A	1.90		23.0	50
RG-213	1.90		21.5	50
RG-214	2.30		21.5	50
9913	N/A		11.0	50
9914	N/A		13.0	50

Cable runs must be considered when installing an earth station because power losses increase with cable length. Most equipment manufacturers recommend using RG-214 or similar cable for relaying signal from the LNA to the downconverter. For transmitting C-band signals over longer distances than normally encountered low-loss, air dielectric cable is required. For even longer distances booster amplifiers are required to recapture the attenuated power. Since cable television companies send their broadcasts over extensive networks, they must use a network of power boosters and splitters even though their signals have been downconverted to the lower-frequency MHz ranges. Note that splitters are devices that divide a signal from one cable into identical messages relayed onto two or more coaxial cables.

The length of cable runs should alway be minimized. This is important because noise, which increases with cable length and decreases with cable diameter, is added to a satellite signal as it journeys from the output of the LNA to the input of the downconverter and from the downconverter to the satellite receiver.

In addition to being chosen accordingly to frequency specifications, coax must be impedance matched to both the low noise amplifier and video receiver. Each brand of coax has a characteristic impedance, a resistance to signal flow, measured in ohms (see Table 4-3). If the output impedance of the LNA does not equal the cable impedance, the signal will "see" this discontinuity and be partially reflected backwards. The performance of an earth station can be completely ruined if the wrong coax is chosen.

In summary, coaxial cable should be carefully selected so that impedances are properly matched and so that their frequency carrying ability is adequate. In addition, distances between the LNA and receiver should be made as short as possible. Also, it is very important that any connectors used must be rated to carry the frequencies in use and that all connections must be secure and waterproof. Many problems in TVRO installations arise from poorly installed cable and connectors.

The Satellite Receiver

The purpose of the satellite receiver is to demodulate or extract the original broadcasted audio and video signals from the carrier waves. When additional stereo information is added to the satellite broadcast signal a stereo demodulator, which is sometimes built-into the receiver, is necessary to demodulate the stereo signal.

Receivers used in the early days of satellite communication were large, clumsy devices whose weight was almost a measure of cost. The modern satellite receiver, which is usually light, small, and attractively packaged, can be compared to the more familiar stereo receiver. A well-designed satellite receiver is reliable, easily maintained, and has high-quality audio and video reproduction (Figure 4-26, 4-27, 4-28 and 4-29).

The downconverter, located at the dish or occasionally built-into the low noise amplifier, is considered to be an integral part of any satellite receiver. This component is often packed and shipped in the same box as the receiver.

How are Channels Selected?

As well as amplifying the weak signal received from a down-converter and demodulating the carrier waves, a satellite receiver selects from among the channels relayed in each downlinked broadcast. Today, most satellites relay as many as 24 channels. A device called a tuner, like the channel selector on a radio or a television, allows any one channel to be viewed by pressing a button or turning a knob on the face of the video receiver. Some commercial receivers do not have this capability but are "semi-agile," whereby channels are selected by changing plug-in circuit boards. Other commercial receivers are fixed to one preselected satellite channel.

Those completely "agile" tuners in consumer brands of satellite receivers are the component most susceptible to breakdowns. Higher-quality tuners "hold" each channel more clearly and are less apt to receive interference from adjacent channels. Synthesized tuners, which are found in all fixed and semi-agile receivers but only in some agile receivers, have this desirable quality. These use a separate crystal (as in a crystal watch) or another component specifically designed to select the frequency of each channel. They are much less

Figure 4-26. New Drake Model ESR924i Component. Incorporating both a satellite earth station receiver and a antenna positioning system in one unit. System uses infrared remote control to select all functions of the 924i unit.

subject to "drifting" than are tuners that use a continuously variable device to tune from channel to channel.

The number of channels that can be viewed at the same time is limited by the number of receivers on-hand. However, the output from an earth station with just one receiver can be detected by any

Figure 4-27. DX-DSB-700 Satellite Receiver with handheld control unit. The DSB-700 can be used on C-Band and Ku-Band Satellites.

Figure 4-28. Chaparral Sierra Satellite Receiver. This receiver has a built-in actuator and a radio controlled hand-held remote. *(Courtesy of Chaparral Communications, Inc.).*

number of television sets. The simplest earth stations use just one receiver, while more sophisticated systems, like those employed by cable TV and SMATV systems (which are small cable systems devoted to one apartment or condo complex), are designed to allow

Figure 4-29. Comtech Receiver. This model commercial receiver has a front panel meter which allows monitoring of signal strength. It can be rack-mounted. *(Courtesy of Comtech Data Corporation)*.

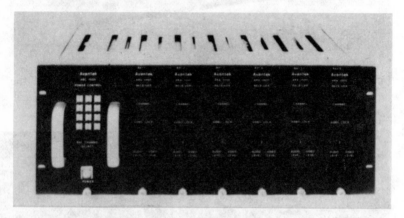

Figure 4-30. Multichannel Satellite Receiver. This unit is really six receivers in one used for commercial, rack-mounted installations. *(Courtesy of Avantek Corporation)*.

many channels to be viewed simultaneously. Figure 4-31 illustrates three possible arrangements that can be used in earth station designs.

What is Threshold Extension?

The threshold of a satellite receiver is a measure of how weak an input signal can be before the television picture quality starts to become unacceptable. Threshold is measured by the ratio of input signal-to-noise power. When the receiver input is adequate—i.e., above threshold—the receiver output signal is linearly related to the input. A linear relationship means that for a given change in input, there is a given change in output. To illustrate, if an input of 1 watt

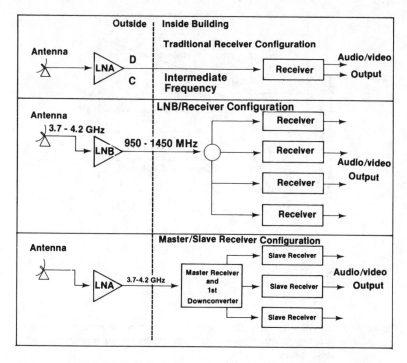

Figure 4-31. Possible Designs for Earth Stations. The bottom two designs are capable of simultaneously receiving more than one television channel.

results in an output of 5 watts, then an input of 2 watts would result in an output of 10 watts. The linear relationship between input and output is important because a TV broadcast of a scene having one area twice as bright as another would not look proper if the reproduced picture had that area only 50% brighter. This could occur if the satellite receiver were not functioning linearly when the input level was below threshold. Many receivers have built-in meters that show how much power is being received from a satellite. These can help determine whether a signal is below threshold.

Most satellite receivers have thresholds in the 7.5 to 9 decibel range (see Figure 4-32); that is, the input signal must be from 5.6 to 7.9 times more powerful than the accompanying noise. As the input level falls below threshold, impulse noise is visible as "sparklies" in the television picture and as random noise in the audio message. This occurs in normal television reception when a signal from a distant station is too weak. Even the best receiver will be incapable of functioning properly if the signal-to-noise ratio generated by the antenna/low noise amplifier/cable is too low.

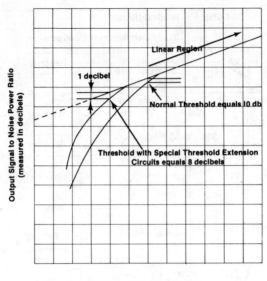

Input Carrier Signal to Noise Power Ratio
(measured in decibels)

Figure 4-32. Satellite Receiver Threshold. This graph of satellite receiver output versus input power to noise ratios shows how threshold is determined.

Threshold extension techniques are used with some satellite receivers in an attempt to lower the threshold in order to improve reception of signals that would otherwise be just below threshold and too weak. Of the three techniques for lowering threshold, using the best possible electrical components that contribute the lowest amounts of noise to a signal processed by a receiver is by far the most effective.

To provide a fuller understanding of the second and third threshold extension techniques, the reason for using frequency modulation in relaying satellite signals must be more clearly explained.

A satellite downlink message when detected by a TVRO is characterized by very low signal power and moderate strength noise power. The frequency modulation technique used in satellite broadcasts purposely spreads the 4.2-megahertz television bandwidth across the available 36-megahertz transponder bandwidth. This use of a wider bandwidth allows for detection of an acceptable-quality TV signal even if the carrier signal-to-noise level ratio is quite low.

The second method used to lower satellite receiver threshold allows only a portion of the bandwidth of the signal from a satellite transponder to be detected in order to reduce noise. This is possible because the amount of noise increases or decreases in direct proportion to the size of the bandwidth. Most of the video information in a satellite broadcast can be recovered in bandwidths as narrow as 14 MHz, substantially lower than the original full 36-megahertz unlinked bandwidth. However, this threshold extension method can occasionally cause some deterioration in picture quality, especially when pictures "loaded" with information are broadcasted.

The third and most sophisticated threshold extension method uses a circuit that not only switches on and reduces bandwidth when the signal is too weak for clear reception, but also "tracks" and centers its reduced bandwidth around the frequency where the television signal is strongest. This technique can substantially lower receiver threshold. However, such threshold extension circuits may cause, as well as solve, problems. If a transponder circuit is occupied with audio subcarriers in addition to a television broadcast, such a circuit may deteriorate rather than improve a weak video signal because not enough video signal power may be contained in the reduced bandwidth. In the worst case, an interfering signal from a source such as a local telephone company may be strong enough so that the threshold extension device would track the wrong signal and completely ruin a broadcast.

Satellite Receiver Quality

The quality of a satellite receiver is judged by both clarity and fidelity of the television picture as well as by the crispness of the associated sound. It must have a low threshold while maintaining a sufficiently wide bandwidth to clearly decipher all the information relayed by each satellite transponder. An inexpensive lower-quality brand might boast a low threshold but this threshold could have been achieved by using a very narrow bandwidth. This may be adequate for most but not all video reception conditions. For example, if a number of audio subcarriers are also present video quality in a "loaded" signal could degrade.

Receiver Signal Power Levels

The original signal sent from a TV camera, the "baseband" signal, has a strength of approximately three thousandths of a watt, 3 milliwatts. This signal has been modulated onto carrier waves, amplified, and attenuated many times, and finally demodulated on its journey to each television set. In spite of the many changes in power of a typical satellite communication link modern technology allows the original message to be clearly reassembled by a satellite receiver.

Modulators and Converters

Modulators and converters are similar pieces of equipment that process the television baseband signal from a satellite receiver into a form that can be detected by a television (Figure 4-33, 4-34, and 4-35). This baseband signal, which contains all the video and audio information as well as the necessary synchronization pulses to recreate a television picture, can be fed directly into a television monitor. A conventional television, however, has built-in demodulation circuits, so the baseband signal must be modulated before being relayed to a TV set.

Figure 4-33. Commercial Modulator. Today most home satellite receivers have built-in modulators. This commercial unit is a separate modulator often used in SMATV and other multichannel installations. *(Courtesy of Comtech Data Corporation).*

Figure 4-34. Commercial Grade Modulator. This crystal controlled modulator has an internal filter to reduce interference between satellite channels. *(Courtesy of Nexus Engineering).*

Figure 4-35. Drake SA-24 Stereo Demodulator. This device allows a TVRO to receive stereo music to addition to video signals. *(Courtesy of R.L. Drake Company)*.

Most home satellite receivers have built-in modulators typically but not always tuned to either channels 3 or 4. Therefore, satellite TV can be viewed by tuning a TV set to either channel 3 or 4. These channels have frequency ranges of 60 to 66 MGz and 66 to 72 MGz respectively. Once the TVRO electronics are hooked up and channel 3 or 4 has been selected on the TV set, satellite TV channels are chosen by tuning the satellite receiver. Note that a home video-tape recorder (a VCR), like a TV monitor, is capable of processing the raw baseband signal. Therefore, signals from a satellite receiver could be fed directly into a VCR whose built-in modulator allows the output signal, to be properly interpreted by a television set.

Converters are generally capable of modulating baseband signals so that they can be received by any chosen VHF or UHF channel on a television. The FCC allocation of these 58 channel frequencies is illustrated in Figure 4-36.

A satellite master antenna system may use a different arrangement for modulating and selecting satellite channels (Figure 4-37). Each receiver is set to a given transponder and has its own modulator tuned to a chosen UHF and television channel. Channels are then selected by the television tuning knob.

A cable company uses another procedure. Broadcasts from each transponder are first modulated into a different frequency band and then relayed via cable lines to each customer's home. A set-top converter (Figure 4-38) selects from the many channels and usually modulates the signals into just one TV channel. Modern cable television systems can be constructed using coaxial cable that can carry up to 58 channels (two lines side by side could then carry 116 channels). The baseband signal is amplitude modulated onto cable lines which feed television sets. Amplitude modulation is chosen because relatively high signal power levels can be used and because the transmission bandwidth must be maintained relatively narrow so that as many channels as possible can be carried by each cable line. This system is different from that used in FM broadcasts where wide bandwidth transmissions of relatively low power are favored.

What are Addressable Converters?

Addressable converters can be either one-way or two-way devices. They are used in cable TV networks and are a parent technology for modern home satellite TV decoders. One-way addressable converters can receive and respond to signals from a cable TV company (Figure 4-39). These small computers can turn subscriber services on and off, keep track of billing, offer once-only special programming only to consumers who wish to receive it, and respond to consumers' needs for changes in program menu with speed and flexibility. For example, if a subscriber wished to drop three channels and add two different ones, a repairman would not be required to visit his home because all changes could be managed from a central location by means of an addressable converter. This form of control

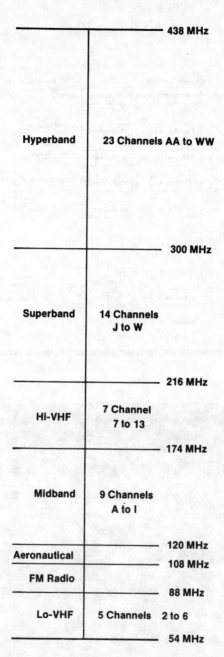

Figure 4-36. Allocated TV Frequency Bands. Each television channel has a 6 MHz bandwidth.

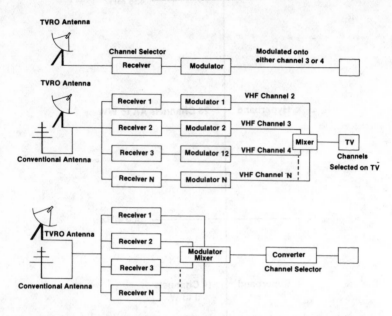

Figure 4-37. Commercial Designs for Modulating Signals onto TV Sets.

Figure 4-38. Cable TV Converter. This 36-channel converter is used by some cable TV subscribers. *(Courtesy of Pioneer Communications of America, Inc.).*

Figure 4-39. One-Way Addressable Terminal and Remote Control. *(Courtesy of Pioneer Communications of America, Inc.).*

is becoming an economic necessity as the complexity of the business increases and many more customers and transactions are being managed.

A two-way addressable converter allows the customer to "talk back" to the cable control center via a home keyboard. Such services can include home entertainment, burglar and fire-alarm protection, medical emergency alert, electronic newspapers, banking, mail, shop-at-home, instant opinion poll voting, videotext, environmental monitoring and control, and energy management.

Scramblers and Decoders

Scramblers are devices that code a broadcasted signal so that the information is not recognizable unless decoded. Such encryption techniques can be used to prevent pirating of cable TV, satellite relays, and many other transmissions of private information. Scramblers can be built into set-top converters so that one-way addressable cable systems can periodically change a code to maintain secrecy. In a similar fashion, satellite transmissions can be coded so that only those who own a decoder and have access to the proper combination will be able to view a broadcast. The Home Box Office

Figure 4-40. Two-Way Interactive Home Terminal. *(Courtesy of Pioneer Communications of America, Inc.)*.

Network has been investigating a scrambling system for satellite TV broadcasts. This type of system would be very secure because a user would be able to send any one of 72 quadrillion combinations it desires over the satellite link and can change the combination at any point in time.

Although, scrambling can be a viable method of deterring all but the most sophisticated pirates, it may not be economically justified in all markets because it may cost more to buy scrambling and decoding devices than the sum of all revenues recovered. The fact remains that improvements in satellite receiving technology have certainly created an unprecedented opportunity for signals to be intercepted at any location under a satellite footprint. Therefore, broadcasters and other communicators are attempting to curtail loss of their revenues by whatever reasonable technical or legal means available.

Figure 4-41. Scrambler/Decoder. The Orion decoder is fully addressable and under computer control by the uplink station. Each unit may be authorized to receive up to 49 different channels. *(Courtesy of Oak Satellite Corporation).*

Build or Purchase Your Own Earth Station?

Some especially talented individuals can build a well-functioning earth station. In fact, in the mid 1970's when the first home earth station was built by a group including a Stanford professor, the cost was comparatively so low that many commercial manufacturers were quite surprised.

If a "do-it-yourselfer" has the time, piecing together a prime-focus or other antenna may be a reasonable undertaking. Antennas can be built from wood, metal screening, aluminum pieces, or other materials if a smooth, true shape can be well approximated. The mounting assembly for the feedhorn and the low noise amplifier must be rigid and must cast as small a shadow as possible onto the dish. It must also be designed so that the feedhorn can be adjusted to the precise center of the dish focal point.

Some talented electrical engineers may even attempt to build the remainder of an earth station. However, the competition between established electronic firms is such that high-quality feedhorns, low noise amplifiers and receivers can be purchased today at prices competitive with those a "do-it-yourselfer" could achieve. Therefore, before an individual decides to build his own system, the comparative cost and the exacting design criteria necessary for a high-quality TVRO station must be carefully considered.

Purchasing and Operating an Earth Station

If the decision is made to purchase a system, one of two routes can be followed to obtain a high-quality system. TVROs can be purchased as a complete or "turnkey" system or each component can be independently selected. Today, components can be bought from many outlets including mail order houses, dealers or individuals who advertise in newspaper classified sections. Whatever route is followed, a consumer should understand the theory of TVRO systems well enough to judge the quality of each component.

It is essential that TVRO consumers should have a practical knowledge of the potential pitfalls enroute to receiving a clear television picture from a satellite. This is just as important as understanding the theory of satellite communication, because, although most dealers are quite competent, some who are selling such systems have a limited and often confused view of the field. It is clear that the satellite TV business is rapidly growing and has yet to see the weeding out of many vendors who sell inferior, poorly matched, or poorly installed systems.

A "site-check," testing the TVRO earth station to be purchased at the planned location of operation, is a crucial first step. If a dealer brings a test satellite system to the planned site, the system should deliver a clear, stable television picture. If reception quality is only marginal during a site-check, then the picture during a rainstorm or other not-so-perfect conditions will be poor. It has been estimated that as many as 10% of all sites in larger cities are subject to a sizeable amount of interference from sources such as the local telephone company. A dealer should be able to suggest alternatives if microwave interference is encountered. For example, if an interfering signal is detected coming from the east, the dish can be moved so it is screened by a tree, it can be placed behind a metal fence, it can be fitted with a reflective shroud around its rim, or it can be moved to a location such as the opposite side of the house. A knowledgeable dealer will also understand the use of filters designed to eliminate interference (please see "The Home Satellite TV Installation and Troubleshooting Guide" by Frank Baylin and Brent Gale for much more detailed information. This book will allow most consumers with previously no experience, to become competent satellite installers and troubleshooters. It will save many hundreds of dollars over the years in "retuning" a satellite system.).

An alternative to bringing in a portable system during a site check is the use of a "spectrum analyser." This instrument is a powerful yet simple method to test for interference and noise at the planned installation location.

All components to be purchased must be designed for the particular location in question. Just because a friend in another city used an 8-foot dish with a similar electronic package does not mean that the system will work elsewhere. An 8-foot antenna may function well in an area near the center of a satellite footprint, but could be inadequate in another location where signals might be more than 50% or 1.8 decibels weaker.

The selected antenna should be of sufficient size to have adequate gain and should have a narrow beamwidth to allow detection of a minimal amount of noise unwanted signals. Remember that antenna gain depends on manufacturing and design quality. For example, a 10-foot dish can have gains that vary from as low as 38 decibels to as high as 40.6 decibels, a difference of 2.6 decibels or 182%! Similarly, low noise amplifiers can have gains that vary from as low as 30 to as high as 60 decibels (a 1000-fold difference in amplification). Although LNA gain is not nearly as important as LNA noise temperature, those having gains less than about 40 decibels should be avoided. Also, if the distance between the LNA and the in-door receiver is long, then a more expensive, lower-noise, higher-gain LNA will be required to compensate for cable losses. The selection and installation of coaxial cables and connectors is similarly important to avoid introduction of excessive noise and signal attenuation.

Many types of receivers are available for purchase today and new models are being constantly improved upon. But remember that "bugs" or engineering problems which often require small design changes need field testing and time to be fully eliminated. It is prudent policy to check the track record of any device before purchase.

Many factors enter into selecting a satellite receiver. They should be of sufficient quality to produce a clear television picture. Lower-threshold units can function better in marginal signal reception situations than can receivers with higher threshold, which are usually less expensive. Brands having synthesized tuning are apt to be more durable than those with continuously variable tuning. Receivers that have built-in, front panel signal power meters allow a dish and feedhorn to be carefully aligned to receive a maximal satellite signal. Some brands are equipped with built-in stereo processors or antenna actuators. In view of all these variations in receiver design it is clear that the needs of each situation must be carefully assessed before a given brand is chosen.

Consumers should be aware of the differences between commercial home-market equipment. Commercial antennas and electronics have been available for longer than 20 years and have, therefore, been carefully tested and improved. Commercial earth stations, however, are more expensive but generally more durable, more carefully designed, and have higher-quality outputs than those sold to the home market. Television station owners cannot afford to lose a signal during marginal reception conditions, and therefore purchase commercial-grade equipment. Unlike the home market where dealers are sometimes in the satellite communication business as a sideline, most commercial dealers are often manufacturers or manufacturers' representatives. It is wise to carefully check on the status of a dealer to determine whether he will be able to service locally installed equipment. Guarantees may only be honored if a dealer is still in business when problems arise.

Once a system is purchased, the equipment must be correctly installed. Polar mounts must be accurately aligned with true north to obtain optimal audio and video. True north is determined by aiming towards the geographic North Pole not towards the magnetic pole as a compass would indicate. The difference between these points, the magnetic variation, is caused by variations in the amount of magnetic materials found in the earth. A compass needle may point more than 40 degrees away from true north in some locations. The correction for magnetic variation may be found from any satellite dealer or surveyor, or by purchasing a magnetic variation map from the U.S. Geological Survey or the local airport.

When the mount, antenna and electronics have been properly installed the "look angles" to find any satellite can be calculated or obtained from graphs of the equations (again the reader is referred to the Home Satellite TV Installation and Troubleshooting Manual). To illustrate, the satellite location chart in Figure 4-42 can be used to find the elevation and azimuth towards each satellite. First, determine the site longitude and latitude and the satellite longitude. The difference between satellite and site longitudes and the site latitude are needed to read this map. For example, if Satcom IV at 83° W is being located from an installation in Houston, Texas having longitude 95.5° W and latitude 29.5° W, the difference of 12.5 degrees and the site latitude indicates that the elevation and azimuth angle needed to target this satellite are 53° and 24°, respectively.

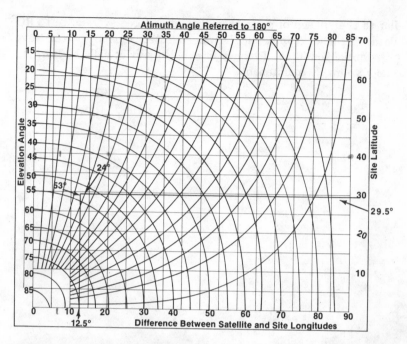

Figure 4-42. Satellite Locating Chart. This chart can be used to determine the elevation and azimuth angles required to aim a dish towards any satellite in the geosynchronous arc.

Once a system is installed and is providing its owner with unmatched entertainment and information, it must be adequately maintained. The amount and type of necessary maintenance is determined in part by the climate and in part by the type of equipment in use. A desert climate will cause more wear due to excessive heat, while maritime conditions could cause more rust and corrosion. For example, LNAs and downconverter may have longer lifetimes if the containers are very well sealed and are perhaps packed with dessicants, materials that absorb any excess water. Probably the most common TVRO failure mode occurs when connections in cable lines are penetrated by water and are shorted out. All cable fittings should therefore be well sealed with a coaxial sealant. Some commercial installations use cables that contain pressurized air between the central conductor and the surrounding metal sheath in part to avoid moisture ingress problems. Other maintenance procedures are dictated by common sense. For example, after a severe wind storm, antenna alignment should be checked. Deviations even as small as $\frac{1}{16}$ of an inch can cause substantial picture degradation. It is important to understand that maintaining a satellite system is more similar to the upkeep of a car than that of a stereo system. A TVRO needs periodic "tuning up." And educated consumers can save substantial amounts of money in avoided service calls by learning "do-it-yourself" maintenance.

Chapter

5

PROGRAMMING

Television as an entertainment medium is changing. As recently as the early 1970's, the three familiar national networks and, to a certain extent, the Public Broadcasting System, produced or controlled most television programming. The majority of UHF and cable TV companies used network-produced shows almost exclusively, although some created their own programming for local distribution.

Today, home satellite systems are an alternative to distribution of programming via either cable TV or conventional over-the-air networks. This has led to unusual, innovative audio and video entertainment and information. As a result, the whole structure of television as an entertainment medium is evolving to radically new forms.

Other Television Broadcasters

Since the early days of television, social, economic, and technological advances have spurred a rapid evolution in the production and distribution of television programming. For example, in the 1940s locally produced and broadcasted TV could not be exported to other urban areas. Therefore, for example, programming originating in New York could not be viewed in Chicago. The installation of cable networks and line-of-sight microwave stations resolved this limitation. During the same period, cable TV companies began distributing network television to smaller cities and towns. Following the introduction of nationwide satellite broadcasting, cable TV companies migrated into the large urban areas with an innovative menu of entertainment and information which was received by earth stations. Today, home satellite systems are a powerful competitive force in distributing satellite programming to both urban and rural areas.

Satellite communications technology allows broadcasts to be received anywhere and subsequently distributed by a variety of methods. The evolving structure of the satellite entertainment industry can be compared to a newsstand, which carries hundreds of newspapers and magazines but which has nothing to do with their writing, production, printing or distribution. In the broadcast business the "writing and production" is the work of independent program producers. The "physical distribution" is the work of companies and engineers who build, launch, service and operate the satellites and ground stations. The "marketing and sales" are the work of cable TV, home satellite TV and other marketing companies.

Special television broadcast modes that have evolved over the years select their menus from the large pool of programming relayed by satellite. In addition to home satellite TVROs, other marketing methods include subscription television (STV), low power television (LPTV), multipoint distribution systems (MDS), cable television, and satellite master antenna television (SMATV).

STV is similar to conventional television broadcasts except that a monthly or per-program fee is charged to customers. The operators of subscription television systems protect their transmissions from unauthorized use by scrambling their broadcasts. Each subscriber's set is furnished with a decoder to allow reception of an unscrambled signal. This type of service has been quite successful over the years. Since the first experimental testing of STV was permitted by the FCC in the early 1950's, at least 500,000 sets have been retrofitted for subscription television.

Low power television, a conventional, over-the-air transmission service, also relays scrambled broadcasts. LPTV is intended to serve the hidden, unserviced, and underserviced rural areas of the country with a combination of local origin and nationally produced entertainment and information. Although cable TV provides excellent service in high population urban areas, the dwelling density in many rural areas is so low that costs to install cables would be prohibitive. Segments of the population living in these areas, as well as groups in large cities not serviced by other broadcasts modes are the target market for LPTV. LPTV has been entitled the "gold rush of the 1980's" because, following granting of the first station permit on May 14, 1982, at least 6500 applications for operating such businesses were received by the FCC within the span of a few months.

The first low power television station, the 17-kilowatt Minnesota ministation, is typical of the breed. It has a 33-mile broadcast radius, which is much smaller than that of a standard 100-kilowatt VHF television station. However, the smaller station can be "turned on" for an investment of just $80,000, which is substantially less than the average outlay of $2 million required for a conventional television station. Because the typical cost for installation of trunk lines for a cable TV distribution system is approximately $14,000 per mile, LPTV is definitely much more economical than installing cable networks in outlying areas.

MDS are microwave broadcasting facilities that rebroadcast satellite programming to individual rooftop antennas on subscribers' residences in the immediate vicinity. In 1984, the FCC has granted permission which allowed multipoint distribution systems to expand their broadcasts from 2 video channels to 31 channels. MDS is economically viable because it reaches an investment breakeven point at only 4,000 to 8,000 subscribers compared to approximately 50,000 subscribers needed for subscription television. In early 1980s there were over 300,000 customers receiving MDS television programming.

Cable TV has been the most successful of all the alternative television distribution methods. Almost all cable television companies now use two or more microwave antennas to receive programming from satellites. Signals from these "headends" are then cabled to subscribers who have a channel selector, and often a decoder, on their television sets. Over 30% of all households in the United States, an estimated 40 million, subscribe to cable TV.

SMATV systems are simply small cable TV networks operating in a mobile home park, apartment, or condominium complex. They are unique in being, in essence, self-contained, mini-cable networks. They generally serve only properties that do not cross public thoroughfares. SMATV systems must use commercial-grade equipment to receive satellite signals and a relatively small cable network to distribute the programming. SMATV systems are not regulated, but a few program producers who scramble their signals still refuse to make their programming commercially available. Nevertheless, such mini-cable systems are firmly entrenched. And the economic viability of SMATV networks is being constantly upgraded as services other than regular TV programming including tenant security, home banking, home shopping, and delivery of newspapers via television are being introduced.

The market potential for SMATV is immense because of the rapid drop in the costs to receive satellite broadcasts. And this technology is quickly gaining a substantial market share. For example, an estimated 25% of cable TV subscribers in Dallas, Texas, now are serviced by SMATV installations. This can be favorably compared to the estimated 20% of the 23 million multiple dwelling units in the

United States that were served by cable TV in 1982. Note that these figures do not include many trailer and mobile home communities that will not be reached by cable networks for many years to come.

Economics of the New Television

Today, consumers can choose from a diverse menu of programming relayed from satellite. This was not the case in the late 1960's when the major television networks owned nearly all aspects of broadcasting including program selection, production, scheduling, and in some cases, transmission. Local TV stations typically broadcasted network programs along with just a scattering of locally produced entertainment. Satellite communications has radically altered the old picture. Today those with innovative financing and creative ideas are now able to broadcast nationally and even internationally from a single uplink station.

But television relayed for mass audience consumption is only one aspect of the market place for satellite communication. Alternative, highly profitable ventures are underway. For example, the American Hospital Video Network links 7000 American hospitals with 24-hour medical education and health-care news. This network produces approximately 30% of its programming and subcontracts the balance to independent producers. Satellite technology is also used for teleconferencing, computer-to-computer relays, and corporate-owned, corporate-used systems. For example, Hewlett Packard owns and operates it own telecommunication system. Telephone expenses alone were reportedly reduced $50,000 per month when the system was first introduced over five years ago.

Video broadcasts directed towards the general public are financed by three methods. Some programmers make their productions available free-of-charge. Religious networks are the prime example of those who seek to relay a message to as wide an audience as possible while expecting no compensation. A second category includes those programmers who also offer their services free-of-charge to the public, but who sell advertising time to commercial sponsors in a fashion similar to the conventional over-the-air networks. For example, the Financial News Network, FNN, is viewed free of charge by over 25 million cable TV subscribers while advertising from hotels, airlines, banks, credit-card companies, commodity brokers, and others who wished to reach a targeted audience pay the bills. A third variety of programmers, the "pay-TV" networks such as Home Box Office and Showtime, charge monthly subscription rates. Periodically, some one-time special events are also sold individually to the public by various broadcasters.

The viewer of the "new TV" is experiencing a metamorphosis in advertising as well as in programming content. Flexible commercial formats such as five-minute, "soft-sell" interludes named "infor-mercials" can target large audiences at relatively low cost. Experts have forecasted that cable television advertising revenues should rise dramatically from $129 million in 1981 to as high as $2.5 billion by 1990. On the other hand, network TV sold $12.7 billion in adver-tising in 1981. This air-time is substantially more expensive than advertising on the "new TV." For example, a 30-second spot com-mercial on an extremely popular network show, such as "Dallas," cost $175,000 in 1981, while the same block of time on Music Televi-sion (MTV) was sold for $1000.

Program Categories

The very wide variety of information and entertainment relayed via satellite can be grouped into a number of general categories. Much more detailed information can be found in the many program-ming guides listed in Appendix C.

News and Weather

News and weather programming, which is often interspersed with general-interest programs such as segments on food or fashion, is provided in a 24-per-day format. An all-day news/weather station has advantages over a network that airs this information for brief periods. Selected news is repeated throughout the day to allow the viewer constant access; as the news changes, updates are released; and as weather conditions change, the viewer is quickly informed.

The pioneer news and weather service was Ted Turner's At-lanta-based Cable News Network (CNN), which was joined by an all news sister station (CNN2) in 1981 (this has since been renamed Cable Headline News). Competition in this area has been keen. For example, a joint effort of Westinghouse Broadcasting (which owns the nationally based Group W news service) and the ABC Network called Satellite Channel News (SCN) was introduced and purchased by a competitor within a span of two years. Another such endeavor was formed when the Eastern Broadcast Service purchased UPI's Newstime to create North American Newstime (NAN). This service relays audio and individual full-color photographs by using slow-scan technology rather than full-motion video. NAN uses a portion of the signal that is relayed to a satellite by WTBS in Atlanta to transmit UPI information. It employs their news-gathering facili-ties to assemble photos, slides, audio reports, and news wires to pri-marily serve newspapers and news offices.

Innovative weather forecasts using sophisticated color computer graphics cover national and local weather developments continuously. One such service available to local cable TV broadcasters is View Weather, which uses a scrambled signal that is delivered along with the WTBS, Atlanta video signal. Operators can choose from a variety of segments including local weather reports, traveler's forecasts, aviation weather, weather history, general climatological data and statistics, agricultural weather reports, and standard national forecasts.

Business and Financial Programming

Business and financial shows concentrate on economic concerns. Two program categories exist. Reuters News Service and the BizNet American Network are examples of pay-only, scrambled business information service. The free, commercially supported ventures such as the Financial News Network (FNN), offer a wider scope of information than the more narrowly-targeted pay-only networks. Business and financial programs relay up-to-the-minute stock and commodity market reports as well as continuous computer graphics and ticker-tape displays. Perspectives on business trends, daily financial news, and reports on many other speculative markets satisfy a more general audience.

Movies

Recently released, uncut films of high quality are offered by some programmers for a monthly fee. Such "premium" programming is among the most appealing to many customers.

Sports Programming

Sporting events such as football, basketball, baseball, tennis, soccer, and hockey supplemented with special-interest stories and coverage are available via satellite. Many sources, such as ESPN, the 24-hour sports cable network, offer these events. USA Cable, as well as superstations WTBS, Atlanta; WGN, Chicago; and WOR, New York, periodically broadcast sporting events. Some local cable stations sell broadcasts of single regional sports happenings to local audiences, while special events broadcast via satellite are occasionally sold nationally on a one-time basis.

Children's Fare

Both educational and entertainment programs targeted exclusively toward young audiences are available via satellite over channels such as the Spanish International Network (SIN), WTBS,

Atlanta, and some religious networks. The cable service, Nickelodeon, is devoted exclusively to the education and entertainment of children. The new programming is in general non-violent and of high quality and has been compared favorably in caliber to the highly regarded "Sesame Street."

Religious Programming

Religious programmers broadcast a variety of events related to organized religious practices and spiritual learning. The programs function both to serve the followers and to raise funds for the churches through donations. In fact, the broadcasts are sufficiently profitable that some cable TV operators have been offered payments to carry programming in order to increase these donations. Networks that do not attempt over-the-air solicitations are the National Christian Network (NCN), the Eternal World Network (EWTN), and the National Jewish Television Network (NJT).

The largest private video downlink network for any use is being built for the Morman Church. This network, comprised of over 500 earth stations, is used primarily for the reception of church communications and functions.

All the religious networks use the miracle of satellite telecommunications.

Adult Entertainment

Programs targeted at an adult-only audience supply all but hard-core pornography. These services include The Playboy Channel and American Exxtasy.™

Music

Music television productions are a relatively recent innovation. Rock, country and other types of music are interspersed with creative video effects and action. These productions are so popular that most newly released albums from well-known artists are now accompanied by video productions. These programs, relayed by networks such as Music TV (MTV) are targeted to the audiences of ages 12 through 34 which comprise in excess of 30% of the national viewing market. Because MTV airs 8 minutes per hour devoted to commercials, this programming is offered free of charge to cable TV operators and the home satellite market. In addition to MTV, a wide range of audio-stereo services, such as the Satellite Music Network and WFMT/Fine Arts Radio, broadcast on sidebands of satellite TV communication circuits. The PBS, ABC, NBC, and CBS network all link their radio stations via satellite.

Family Programming

Family programs are composed exclusively of wholesome material that excludes sex and violence. For example, the Home Theatre Network carries family shows and is available as in a pay television format 12 hours per day. Four of the twelve hours on this network are devoted to travel programs, with the remaining time given to drama, comedy, musicals, adventure, and children's films.

Special Interest

Special-interest programs cater to a clearly defined, narrow market segment. For example, the Spanish International Network and its venture, Univision, is geared to the Spanish culture, while Daytime produces programming aimed specifically at the American Woman. As a joint venture of ABC Video Service and the Hearst Corporation (publisher of such magazines for women as Good Housekeeping, Cosmopolitan, Harper's Bazaar, House Beautiful, and Town and Country), Daytime produces broadcasting in areas such as nutrition, health, cooking, relationships, and parenting. Another special-interest network, the Black Entertainment Network, targets the interests of black Americans. The American Satellite Network and Select TV transmits movies to hotel chains only.

Political Programming

The network C-Span is publicly owned and broadcasts directly from the U.S. Congress. It relays coverage of many public events in the nation's capital in addition to monitoring the U.S. Congress and Senate. It is both a media for politicians and a source of information for the discriminating electorate.

Educational

An example, of educational broadcasts is the American Educational Television Network. It has an audience of approximately 30 million people who receive credits for "going to school" via TV. Most viewers are professionals who must periodically update their knowledge. The network charges relatively high rates for "informercials" that are directed at these highly targeted audiences.

Programmers and Satellite Assignments

Table 5-1 lists the satellites designed for transmission of video broadcasts to the United States and Canada. As the number of transponders are increasing, so are the number of television shows. All programming is assigned to one or more transponders or relays aboard the telecommunication satellite. Each transponder is allocated a given microwave frequency and polarization so that an earth station may tune to any desired channel by choosing a satellite, a frequency band, and a feedhorn polarization.

Orbital Location (°W)	Early 1982	Mid-1092	Early 1983	Mid-1983	Early-1986
			TABLE 5-1. VIDEO SATELLITES		
143	Satcom F1		Satcom F5	Satcom F5	Satcom F5
139			Satcom F1R	Satcom F1R	Satcom F1R
134				Galaxy I	Galaxy I
131	Satcom F3R	Satcom F3R	Satcom F3R	Satcom F3R	Satcom F3R
127		Comstar D4	Comstar D4	Comstar D4	
125					Telstar 303
123	Westar II	Westar V	Westar V	Westar V	
122.5					Westar V
120					Spacenet I
119	Satcom F2	Satcom F2	Satcom F2	Satcom F2	
114	Anik A2/A3	Anik A2/A3	Anik A2/A3	Anik A3	
113.5					Morelos F1
112				Anik A3	Anik A3
109	Anik B	Anik B	Anik B	Anik B	Anik B
104.5					Anik D1
104		Anik D1	Anik D1	Anik D1	
99	Westar I	Westar I	Westar IV	Westar IV	Westar IV
96					Telstar 301
95	Comstar D2	Comstar D2	Comstar D2	Comstar D2	
93.5					Galaxy 3
91	Westar III	Westar III	Westar III	Westar III	Westar III
87	Comstar D3	Comstar D3	Comstar D3	Comstar D3	Telstar 302
84					Satcom F4
83	Satcom IV	Satcom IV	Satcom IV	Satcom IV	
79	Westar I/II	Westar I/II	Westar I/II	Westar I/II	Westar II
76					Comstar D3/D4
74					Galaxy 2
72				Satcom F2R	Satcom F2R
70				Spacenet II	Spacenet II
69					

Chapter

6

IS IT LEGAL?

Is it legal? This seemingly simple question has arisen regarding many of the new technologies, markets, and business forms in the growing telecommunication field. The proliferation of low-cost earth receiving stations has been accompanied by questions of "freedom of the airways" and has been clouded with legal battles about "pirating." Important questions about who may own satellite communication channels have also been raised. The new technology and the potential for enormous profits have encouraged many investors to enter the communication business and has blurred the distinction between the responsibilities and regulation of the participants. These are among many that issues that often lead to legal confrontations and re-evaluation of the telecommunication business.

In the past ten years changes have occurred so rapidly that the previously enacted legislation, the 1934 Communication Act, has become very much outdated. In the process of writing new legislation, the U.S. House of Representatives Subcommittee on Telecommunications conducted a comprehensive examination of telecommunication policy, which included the following:

"An examination of the effects and status of competition and deregulation; an analysis of the electronics and telecommunications industries and their role in world trade; an investigation of Western Hemisphere and International issues; a look at the panoply of emerging home and business information technologies; scrutiny of the role that the federal government plays in telecommunications policy development, including federal, state, and local jurisdictional disputes . . . the possibility of deregulation in international telecommunications services, the reaction of Third World countries to the development of a global information system, concerns over privacy and trans-national electronic mail, the future roles of Comsat and Intelsat in international communications."

Pirating or Right?

The most hotly debated legal issue has centered around the home satellite TV business. The labelling of owners of TVROs as "pirates" and as "thieves" in the earlier days of home satellite TV had as its basis the debate about economic rights of programmers, program distributors, and consumers. Many consumers have argued for "freedom of the airways." For example, when private reception of satellite broadcasts from American programmers was outlawed in Canada by its federal government, the right to free airways was forcefully stated by a Canadian politician: "The law is an ass, if that is the law . . . If some body wants to put a signal down in my backyard, nobody can tell me not to look at it."

It is clear that programmers who incur considerable expenses in producing and broadcasting television to the public deserve a just return on their investments. If too many consumers received pay-television signals free-of-charge, programmers would either suffer financial losses or be forced to sell commercial time. The legal questions relating to private reception of satellite signals are further complicated because many other broadcasts are offered free-of-charge by religious or non-profit groups or by commercially sponsored networks.

The legal right of individuals to receive satellite signals has depended upon interpretations of both copyright and communication laws. The copyright laws, which were updated in 1976 to include video and cinematic entertainment, protected producers who requested payment from their customers. Also, while Section 605 of the 1934 Communication Act prohibited the "interception and unauthorized use" of private television programs over public airways, one interpretation did suggest that those who use but do not resell transmissions were within bounds of the law.

The battle over the legality of home satellite TV became heated after the FCC 1979 deregulation ruling which eliminated license requirements for earth stations that only receive but do not uplink satellite signals. Following this landmark decision, programmers and the cable TV companies repeatedly attempted to outlaw the home satellite TV industry. In the early fall of 1980, a Congressional amendment, the "theft of service" legislation, prohibiting private reception of satellite signals by thousands of TVROs, was defeated on the grounds that this law would hamper development of innovative satellite technology. The lobbying organization for owners of home earth stations, the Society for Private and Commercial Earth Stations (SPACE), which spearheaded the defeat of this amendment and is still a powerful force in the industry, argued that three classes of citizens who watch TV had been created. They are "those who have paid their bills, those who haven't, and those who aren't allowed to." Although there are numerous instances of private earth station operators offering to pay for their services, HBO as well as other pay TV programmers had returned subscription payments citing either contracts that limited them to servicing only the cable television industry or a lack of policy on dealing with individuals. The weak argument often used was that collecting small monthly bills from thousands of individual subscribers scattered around the country would cost more than the revenues received.

SPACE and the National Satellite Cable Association (NSCA), an association formed to represent owners of small "mom and pop" cable TV systems, have argued that people in rural areas should have the same program-viewing opportunities as the urban American. They claim that many rural Americans cannot purchase cable TV or television from other supplies such as multipoint distribution systems or subscription TV services at any price. Today, an estimated 10 million Americans have either no or marginal television reception without the use of a home satellite system. However, the rapidly growing home satellite business is radically changing this situation.

Many people instilled with a pioneering spirit have insisted on freedom of the airways and the right to receive signals directly from satellites. Most of these "pioneers" were willing to pay the programmers for their services. SPACE had adopted a consistent and fair twofold policy: "Every individual has the right to receive any information transmitted through the air; and that information intended to be paid for must be paid for—not stolen." The Canadian ban on receiving American satellite broadcasts had not deterred an estimated seven hundred private users from illegally installing TVROs and is a good illustration of how this right was perceived. A prominent Canadian politician advised his Federal government, "If a signal is falling on your property and you don't want it, ask them (the Americans) to take it off. If you want to compete, then turn out better programs." This independence of attitude caused the Canadian government in March of 1983 to lift the ban on use of private earth stations. Too many home TVRO owners had simply ignored the prohibition.

By the mid 1980s the heart of this issue had become clear. It centered around who would have the right to market and distribute programming. Programmers were naturally more willing to sell their broadcasts in bulk to parties who serviced many customers than to individuals. They also felt an allegiance to the cable industry which had been responsible for most of their growth and revenues. Both the cable and programmers enterprises had grown in step during the 1970s and early 1980s. But the cable industry was attempting to gain a monopolistic control over the distribution of satellite television.

The situation had begun to change in 1983 and 1984 when "minicable" companies, the master antenna television (SMATV) businesses, were in the process of capturing an increasingly large share of the cable TV market. They fought long, hard and ultimately successfully in obtaining the right to purchase quality programming. For example, some producers of "premium programming" who refused to sell their product to SMATV companies had been sued in 1982 by the State of Arizona for violating anti-trust laws. The NSCA had successfully argued that a number of large, franchised cable operators were pressuring their affiliated program suppliers in an attempt to keep the small operators of cable and SMATV out of the business.

In contrast, bars and restaurants equipped with satellite receiving stations were never allowed to purchase premium programming. While they could legally air free programming such as Music Television, pay television was simply not available. These programmers wanted to sell individually to all the bar patrons via home-based cable television networks and thus earn substantially higher revenues than by selling once to each bar. In fact, HBO had systematically and successfully sued bar owners for theft of their signals and the convicted owners were often required to pay them a penalty of three months worth of revenues.

On October 30, 1984, satellite TV was formally legalized by a rider attached to the Cable Communications Policy Act, which amended Section 605 of the 1934 Federal Communications Act. It clearly stated that it is legal for individual home owners to intercept and view any unscrambled satellite programming and permitted satellite programmers to market nonscrambled broadcasts only if a "good faith" marketing plan had resulted from fair marketplace negotiations. But it increased civil and established criminal penalties for illegal use of satellite signals including the unauthorized viewing of satellite signals in public places. It also left the major issue of access by home users to scrambled signals unresolved.

However, the development of reliable scrambling and decoding devices has finally allowed a technical resolution to the issue of reimbursing pay TV programmers. Television broadcasts can be protected, at considerable expense, by incorporating these sophisticated scrambling and decoding devices into television distribution systems. "Smart" addressable decoders are being used by some broadcasters so that codes can be periodically changed to stop reception of pay TV programs.

The M/A COM Videocipher system developed under contract with HBO is a good example of the capabilities of scrambling/decoding technology. Each individual decoder has a built-in unique address which allows it to be turned on and off at will from a central control site. This gives the programmer the flexibility to offer one-time, pay-per-view special events. Codes can be changed so often that security is well maintained. Although the video quality is somewhat degraded by the coding/decoding process, it is argued that audio enhancers actually improve sound clarity. The decoder displays any type of message as well as diagnostics directly on the television screen. M/A Com can also license the technology to be built-into a standard satellite TV receiver.

But the legal and institutional resolution to these problems has lagged considerably behind the technical solution. The central issue is the development of a fair and equitable method for all parties to be able to receive satellite broadcasts. And the debate has pitted the home satellite industry against the cable TV giant.

Economic rights of both the cable and home satellite TV industry must be protected. The typical cable TV company incurs considerable expense in "wiring" a city. This plus the system operating cost is then passed onto the consumer. For example, each premium channel such as HBO, Cinemax or Showtime typically charges $10 per month per household. Cable TV companies pay a wholesale price of approximately $4 per month with the difference covering their expenses plus their profits. In 1985, following extreme pressure from consumer groups and the satellite industry, HBO announced that it would make its services available directly to the home satellite owner for $12 monthly. This charge did not include the required purchase of a $395 Videocypher decoder. This was perceived by many as unfair since the owner of a TVRO has already invested $1000 to $4000 in the capital cost of a TVRO and should not be asked to pay substantially more than the wholesale cost plus a reasonable extra for billing expenses. In effect, the TVRO owner had become a micro-cable company and expected to receive near wholesale rates.

Today, some experts feel that the writing is on the wall! It is judged that satellite TV has greater flexibility than cable TV and other distribution methods. It offers more choice, reaches even the most remote rural regions, does not require expensive, high maintenance cabling systems, creates more jobs and allows more personal freedom than cable TV. Cable networks certainly have their market niche for some time to come in the higher density urban areas. But as satellite TV grows and matures, new technologies will eventually begin to erode even thoses strongholds. City-wide cable TV networks may even be an interim and relatively short-lived technology. This expectation may be the reason for their fierce battle to retain an almost monopolistic control over distribution of satellite television programming. However, some cable TV companies have taken a different tack and have wisely entered the business of selling or leasing home satellite systems.

It is interesting to view this discussion from a slightly different perspective. It has been estimated that from 10 to 20% of all cable TV transmissions are illegally intercepted, resulting in losses of many millions of dollars of revenues per month. Since nearly 40 million households subscribe to cable TV, this number ranges from 4 to 8 million unauthorized viewers. Even at the end of 1985 only approximately 1.7 million homeowners had installed TVROs, less than half the number cable TV thefts. Even scrambling cable TV signals by the proposed method does not cure this ill since decoding occurs at the headend and "clear" signals are available on the entire distribution network. The only effective solution to maintaining control over satellite programming is to decode the signal at individual subscribers' homes, a solution which is a natural to the home satellite TV industry.

In spite of the uncertainty surrounding this debate, owners of satellite TV systems have an enormous selection of over one hundred video channels. And those who wish to purchase pay TV are comforted with the knowledge that they are financially supporting the creation of quality entertainment. The day is near when "program brokers" will offer a selection of five to ten pay TV channels for a reasonable fee which will be selected via an 800-number. And all these broadcasts will be decoded by electronics built into satellite receivers.

Other Distribution Services

Answers to legal questions about pirating of subscription television (STV), multipoint distribution system (MDS) broadcasts, and low-power television (LPTV) were seemingly more complicated because these services are not relayed from satellites and are also subject to a "free-airways" interpretation. STV signals have been scrambled in codes so simple that a host of companies, especially in

California, were founded just to sell decoders to "pirates." Although these companies have been acquitted in a number of law suits aimed at curtailing their sales, some state and municipal governments have passed legislation outlawing their activities. Also, Section 605 of the 1934 Communication Act has been successfully used to outlaw unauthorized reception of MDS or STV on the ground that these broadcasts were not intended "for the use of the general public." Nevertheless, although numerous injunctions have been issued to stop sales of equipment to receive MDS broadcasts, many thousands of homes still have the small microwave dish and electronics necessary to intercept these programs for free. In June of 1982, HBO, the major MDS broadcaster, first successfully sued an individual in Minneapolis for pirating. Before this action only equipment manufacturers were sued. The recent introduction of LPTV stations, which will also scramble broadcasts, opens up a new market for pirates and manufacturers of decoding equipment.

The Right to Satellite Circuits

Until the mid 1970's the distinction between broadcasters and "common carriers" was clear. Common carriers transmitted audio and video signals from point-to-point over relatively long distances for their customers. And broadcasters had distributed audio and video messages from one point to a local market at no cost to the end user. The roles of common carriers and broadcasters has become less clearly defined because satellite communication allows a wide range of business structures and technical solutions to a given communication need. As a result, the applicability of the 1934 Communication Act has been questioned and new legislation was studied.

The FCC regulates the use and ownership of transponder circuits in its effort to protect the public interest as mandated in the 1934 Communication Act. In the 1970's the FCC required that satellite operators meet technical and economic requirements as well as serve the public. However, applicants were not expected to have a common-carrier status as a condition for launching and operating a communication satellite. The demand for transponder circuits was sluggish until the early 1980's, so users simply leased these circuits. As the demand grew, sale of the full-time use of a transponder to the highest bidder became a viable business alternative. In an effort to be as free from FCC regulations as possible, operators argued that they were not common carriers, since the transfer of a transponder to a user is not an offer of a service for hire to the public. To further support the argument that they should not be governed by the existing communication legislation, they contended that the sale of a transponder was not really a sale, but a pre-paid, protected lease, and that satellite owners actually retain control over maintenance and operation of the complex vehicles.

Sales of transponders began in the early 1980's, partially as a result of the success of these arguments, and partially as a result of the FCC policy to allow a competitive industry to emerge. On November 12, 1980, the Western Union Telegraph Company informed the FCC that it had completed a contract with Citicorp, a large New York bank, that included an option to purchase a transponder aboard the Westar satellite. On December 11, 1980, Citicorp exercised this option. On March 11, 1981, Hughes Communication Incorporated informed the FCC of its intent to sell transponders on its newly authorized Galaxy space vehicles. Then in November, 1981, RCA Americom auctioned ownership of transponders onboard its Satcom IV satellite. They claimed that this action was motivated by a desire to realize a return based on market value, not on cost. This auction would have earned RCA $90.1 million for the sale of seven transponders. One company had bid as high as $14.4 million for rights to a single transponder! But three months later the FCC ruled that the process was unfair and unacceptable. Many of those who were unable to purchase or lease time on Satcom IV took RCA to court for redress of lost cable TV revenues.

The method of allocating satellite communication circuits has been a subject of considerable debate. Operators argue that by selling transponders and by realizing a higher return on the expensive business of building, launching and operating a satellite, they would encourage a rapid expansion in satellite capacity. In fact, by granting approval to launch an additional twenty satellites at the beginning of the 1980s, the FCC had an avowed policy of creating such a competitive and profitable environment. Transponder users argued that the purchase of communication circuits would permit improved business planning. Others had contended that sales would squeeze out the smaller, less richly financed entrants in the telecommunication business who could not afford the market prices but who might have been able to bear regulated prices. It is possible that the number of satellites and thus satellite transponders will outpace demand in the latter part of the 1980 so that some of these problems will be more manageable. In any case, the FCC approved sale of transponders on a case-by-case basis in August 1982. Today the vendors of these satellite circuits are not considered common carriers.

Regulation or Deregulation

Both the broadcast industry and the common carriers have been carefully regulated over the years. In the early, simpler days of radio and television, the FCC was called upon to control only over-the-air broadcasts. Regulation of the giant common carriers was also less complicated because the services offered by AT&T and a few competitors were more clearly defined.

The proliferation of innovative broadcasters and the development of satellite communication technology has raised many questions. For example, will low power television aimed at serving special interest groups interfere with cable TV or network television signals in metropolitan areas? Should the FCC even attempt to regulate competition between subscription television, multipoint distribution systems, cable TV and the video cassette industry, all of which are aimed at the urban consumer? Are two way addressable home communication systems the business of cable operators, the local telephone company or other businesses?

Since the legislated breakup of AT&T, this company as well as other large common carriers are now permitted to enter all aspects of the communications business. Many regulatory issues have been raised by the new business arrangements. For example, how well can some of the new long-line telephone services compete in a deregulated market with AT&T which owns essentially all of the terrestrial long distance networks, is establishing its own satellite communication system and has an operating budget larger than that of most nations of the world? Newspaper publishers who rely heavily on advertising fear that AT&T's could create an "Electronic Yellow Pages" that could significantly decrease their advertising revenues.

Many questions have been raised here. It is clear, that only the future can provide definite answers to these complicated questions.

Chapter

7

THE FUTURE

The satellite communication field is continually expanding and evolving. New technologies and uses for these technologies are constantly being discovered. Today, novel forms of television such as the higher frequency direct broadcasting systems as well as cable TV and C-band home satellite systems are already causing revolutionary changes in the television industry. Other technologies such as videoconferencing, corporate data and telephone networks and a host of related systems are tying our globe even closer together. As a result, many established as well as entrepreneurial businesses have and will continue to experience rapid growth.

These developments have led to an increased demand for satellite transponder circuits and to hightened competition for the use of space in the geosynchronous orbit. Today, more than ever before, a clear need exists for international cooperation in allocating orbital space and in developing technologies that increase the available communication capacity.

Direct Broadcast Systems

The relatively new Ku-band technology named direct broadcast systems (DBS) which employs 2 to 3 foot dishes for receiving satellite television has tremendous promise. Some predict that millions of these small rooftop antennas could, in time, even replace conventional over-the-air television.

DBS is based upon years of technical developments. The joint American/Canadian Communications Technology Satellite (CTS) experiments in the latter half of the 1970's demonstrated that using higher-frequency microwaves for relaying signals to relatively small antennas was feasible. As a result, the 1979 World Administrative Radio Conference (WARC) placed a 154 member nation stamp of approval on DBS. Technical standards set at this meeting called for use of extremely powerful, 200-watt satellite transponders operating in the 12.2 to 12.7 frequency band. In comparison, most C-band satellite circuits presently operate at under 10-watts. This planning group envisaged use of very small 1 to 2 foot dishes for receiving DBS signals.

In 1982, Canada took the lead in implementing the first limited-scale, but fully operational, DBS system broadcasting to remote rural locations. In the United States, the FCC, in response to interest expressed by the private sector, requested and received by the July 16, 1981 deadline proposals from fourteen businesses who predicted that they would design, build and operate direct broadcast television systems by as early as 1985.

How Does DBS Work?

DBS systems are different from the more familiar C-band direct satellite broadcasts because they can potentially use a much smaller, more manageable antenna. A number of satellites are now relaying higher-frequency 14/12 gigahertz transmissions to antennas ranging from 2-1/2 to 6 feet in diameter. Dishes smaller than those required for conventional, lower-frequency, 6/4 gigahertz broadcasts can be used because Ku-band downlink antennas have more narrow beams width and therefore are capable of having more restricted footprints and higher EIRPs. In addition, each transponder aboard a DBS satellite today is capable of generating nearly 50 watts of power, which is 5 to 7 times greater than that of transponders relaying C-band broadcasts.

When the concept of DBS was introduced it was met with strong interest. It had been "certified" at the 1979 WARC and it also included the concept of using scrambled broadcasts which provide programmers a mechanism for collecting revenues. Subscribers would pay a monthly sum to cover the programmers fee plus a portion for leasing the necessary reception equipment. In some cases, customers would be expected to purchase the required antenna and the electronics to decode the satellite signal. In contrast, C-band home satellite systems were regarded in the earlier days of satellite television as an uncontrolled nuisance because this technology had grown unexpectedly as a "grass roots" movement and no mechanism had been incorporated to collect revenues from individual TVRO owners.

The DBS pioneers realized that their profitability depended upon the availability of low cost equipment for receiving satellite transmissions. It was predicted that the necessary antennas and associated electronics could be mass-produced for as low as $200 or $300 per system by as early as 1986. Comparable costs in 1982 were as high as $1000. This rapid drop in cost was predicted to give DBS the unique advantage of broadcasting affordable entertainment via satellite to a national audience.

There were initially some criticisms of DBS. It was predicted that the Ku-band signals would interfere with some earth-based communication systems and therefore cause economic hardship. For example, it has been argued that interference from DBS broadcasts in the 12-gigahertz band, also used in Los Angeles for law enforcement, fire protection, paramedic units, computers, and mobile radio unit communications, could cost the city $15 million in purchasing new compatible equipment. Some expected that multipoint distribution system television operating in an adjacent frequency band, might also experience some similar interference problems.

The Success of DBS

The success of any form of broadcast television depends upon how conveniently and at what cost quality entertainment and information can be viewed. The consumer, in general, does not care which type of program delivery system is used.

Some have forecasted that DBS technologies would be the death of cable television and conventional over-the-air broadcasts. These predictions may parallel similar unreasonable expectations in the past that television would bankrupt radio and that cable television would kill conventional television. A more realistic view is that each alternative mode of broadcasting will capture a share of the television viewing market, which itself has many specialized components. For example, DBS and conventional home earth receiving stations face no competition in far-removed rural areas, while cable television is most profitable in large, densely populated cities. It is important to realize that this balance could easily be upset unless each broadcasting mode is controlled by fair and equitable regulations.

The growth of direct broadcast systems has and is being influenced by many forces. The C-band home satellite television industry has grown astonishingly quickly, much to the amazement of many industry planners and specialists. It in itself has become a "direct broadcast" technology which rivals DBS especially as adequate reception can be attained with dishes as small as 6 feet in diameter in central regions of the United States. The selection of programming available via C-band broadcasts also shadows that which had been planned for DBS.

However, DBS broadcasts are beginning to capture market share in a form different from that originally foreseen. First, it is clear that the selection of programming as well as the cost of earth station terminals will ultimately determine the potential DBS market share. In order to be competitive complete receiving systems must be available for less than $600 to $800. Second, the DBS operators who implement well-planned services more quickly, will fare more successfully in competition with cable TV, low power television (LPTV) and C-band home satellite systems which are all rapidly expanding into receptive markets. Certainly these television distribution industries are vying for similar markets. For example, even though DBS has a unique advantage in remote areas, the growth of LPTV in more densely populated rural areas and C-band home satellite transmissions in most markets areas offers stiff competition.

The outcome of the rivalry between home satellite systems, DBS, LPTV, cable TV and conventional TV for new markets is hard to predict. It is clear that a significant need for new broadcast forms does exist because, in early 1982, in excess of one million American homes had no television service and 2.3 million had only one available channel. It is estimated that over 10 million residences today have either poor reception of a limited number of channels or no reception at all. Note that all these competing distribution methods have access to similar programming relayed via satellite. But some like LPTV have the additional strength of being able to offer locally produced entertainment and information.

The issue of how advertising dollars are spent will also affect the market competition. Programmers' decisions to advertise with any particular broadcast group are driven by what return they realize on their invested funds. For example, most direct broadcast systems would be enormously successful if they had 3 to 5 million subscribers. But why would an advertiser who could easily reach 10 million residences via cable or conventional television limit the targeted audience to 3 or 4 million unless air time was substantially less expensive? The National Association of Broadcasters, the lobbying arm of conventional network TV, has expressed the worry that LPTV and network television relaying commercially sponsored local programming might not be able to survive on local advertising money alone if much of the lucrative nationally directed advertising was sold exclusively to national broadcasters such as DBS and cable television. However, some perceive that the strength of DBS lies not in commercially sponsored but pay TV entertainment.

Cooperation rather than competition between DBS and other broadcast modes is possible and probably advantageous for many.

DBS broadcasts can be relayed into cable TV, LPTV, or satellite master antenna systems for distribution. Cable TV companies, which are already organized to service millions of subscribers, might be in an ideal position to install and service DBS or more conventional C-band satellite equipment. It is important to note that the organizations which can best service the new high technology equipment certainly have a competitive edge.

The Would-Be DBS Operators?

In 1981 Comsat filed a massive proposal with the FCC to provide three channels of DBS service to an estimated 35 million homes. A separate entity, the Satellite Television Corporation (STC) was formed to own and operate the system of 6 proposed satellites, 4 to span the entire United States and 2 to remain as in-orbit spares. By the July 16, 1981, FCC deadline, thirteen other companies had submitted detailed proposals. Three additional firms, United Satellite TV, Oak Satellite Corporation, and Satellite Syndicated Systems, Inc., also filed for DBS authorizations after the cut-off date. The DBS Corporation proposed to operate as a DBS carrier offering free TV to subscribers but charging programmers as much as $1000 per hour for leasing each broadcast channel. Other companies also had unique aspects to their plans. Western Union envisaged using the 11.7 to 12.2 GHz instead of the 12.2 to 12.7 GHz band proposed by all other participants. Focus Broadcasting planned to lease transponder time on Western Union's advanced Westar satellite to make experimental services available one to two years before competitors and to serve only Miami, New York, Los Angeles, San Francisco, and selected rural areas. CBS proposed using a new form of digitally transmitted broadcasts named "high definition television," which would result in a much clearer picture than conventional TV.

Other countries have also attempted to implement direct broadcast systems. Canada has operated a limited-scale DBS system since 1981. The United Satellite Corporation, a subsidiary of British Aerospace, GEC-Marconi, and British Telcom plans to introduce a DBS system via the Britsat satellite. They expect to broadcast two channels to 2 1/2-foot antennas on the British Isles and to dishes as large as 9 feet on the European continent, which would be in a weaker footprint area. The Japanese, among the leaders in high-frequency satellite broadcasts, have had Ku-band experiments underway for many years. Italy, France, and Germany are also actively involved in their own DBS systems.

The first DBS service in the United States was successfully initiated by United Satellite Communications, Inc. (USCI) in late 1983 reaching Indianapolis and 34 counties with five available channels. This firm planned to offer DBS programming by late 1984 in those 26 states covered by the satellite Anik C2. The failure of this venture in 1984 and the withdrawal or retrenching of other companies who had planned to introduce DBS to the American market sheds light on the future of DBS.

It is now clear that high quality reception of Ku-band signals from 50-watt transponders using dishes as small as 1 or 2 feet is a reasonable expectation. So many of the original DBS schemes failed because the technology was simply not ready but had been simply mandated "by committee." The hope of having 200-watt transponders simply was not realized in the expected time frame. The home satellite industry which has so successfully introduced C-band direct broadcast technology is now involved in pioneering workable Ku-band systems. In addition, a significant market for dual-band systems has recently developed. Today the future of Ku-band DBS systems looks bright because the introduction of these technologies is following a market-driven path.

The Battle for the Orbital Arc

The rapid growth and development of satellite communication has sparked the interest of nations around the world. Satellites populating the geosynchronous equatorial arc can simultaneously serve many countries and are therefore the concern of all mankind. However, only so many microwave relays can orbit the earth at once before severe problems with interference between satellite transmissions will arise. As a result many institutional and technical solutions are being incorporated into the global communication system to allow the use of a maximum amount of microwave relay circuits and to foster global cooperation.

Global Harmony?

Satellite services by their very nature transcend national boundaries. Since the United States and Canada are not the only countries launching telecommunication satellites into the equatorial arc, they must consider the concerns of many neighbors. American satellites and the Canadian series of Anik spacecraft occupy the same arc as Intelsat, Brazilsat and Mexican Morelos vehicles. Other South American countries including Argentina and Columbia are also planning to launch their own satellites. When Intelsat was formed to regulate, own and manage the international telecommunications network it did not envisage the technical developments that now permit operation of privately owned international satellite communication systems. Even the 1962 Communications Satellite Act which marked the entrance of the United States into the Intelsat Organization was vaguely worded. It stated that "... it is not the intent of Congress to preclude ... the creation of additional communication satellite systems, if otherwise required in the national interest."

In October 1981 the FCC approved 13 applications allowing U.S. domestic satellite systems to transmit signals across international borders. Seven of these petitions received by the FCC requested permission to receive data from Canada; two proposals, one filed by Satellite Business Systems and one by the American Satellite Corporation, sought to transmit to as well as receive data from Canada. Four companies requested to transmit or receive television programming to or from points in Canada, the Caribbean, and Central America. In early 1983, 15 more applications for transborder service between the U.S. and Canada, Mexico, the Caribbean and Central America were approved. These actions questioned the role of Intelsat in international communications. This was the case even though the the original Intelsat treaty required approval by all member countries as well as coordination by Intelsat to ensure technical compatibility with the Intelsat system and to avoid significant economic harm to the overall system. When Intelsat was created all 154 member nations were expected, except in rare instances, to use the Intelsat facilities. However, modern satellite communication technologies have become reasonablly priced and permit extremely flexible system designs. Today, some perceive that the issues of commercial competition, and national self-interest have become as important as the need for organization to prevent interference and to foster cooperation.

International meetings, such as the 1983 Region II Administrative Radio Conference (regional WARC) and the 1985 and 1987 World Administrative Radio Conferences (WARC) sponsored by Intelsat, have been organized to resolve issues relating to frequency band allocations, international communication networks, and satellite spacing and location. The burgeoning demand for transponder circuits in the United States alone had been more than sufficient reason for careful planning. The issues are even further complicated because other neighboring countries use the same portion of the geosynchronous arc.

The international community can choose some technical remedies to increase the number of available interference-free transponder circuits. These fall into the general categories of space and bandwidth conservation. If satellites relaying as many channels as technically possible and operating in multiple frequency bands were spaced more closely together, more transponder circuits could be made available. More efficient use of the available in-orbit circuits can also be achieved by direct inter-satellite communication, by launching spacecraft which are capable of switching between multiple beam antennas, and by using techniques to squeeze as much information as possible into each portion of the allocated bandwidth.

Satellite Orbital Spacing

Spacing satellite closer together increases the potential number of available transponders, but may result in problems with both uplink and downlink interference. Without any new technical fixes, decreasing spacing forces designers to face a trade-off between the extra cost of ground-station equipment and a need to increase the number of available satellite circuits. As satellites are spaced more closely together larger, more expensive antennas which have more narrow beamwidths must be used to avoid interference and a deterioration of signal quality.

The issue of satellite spacing had not been seriously considered until 1980. RCA Americom had been operating two satellites, Satcom II at 119° W and Satcom I at 135° W, while Western Union had three spacecraft at 91° W, 99° W and 123.5° W. In addition, a joint venture of AT&T and GTE had leased the entire capacity of Comstar General's fleet, Comstar I at 128° W, Comstar II at 95° W and Comstar III at 87° W. A tentative policy had allocated 6/4 GHz, C-band and 14/12 GHz, Ku-band satellites to a minimum of 4 and 2 degree spacing, respectively. In October, 1981, the FCC surprised the industry by requesting public comment on the possibility of authorizing spacing of satellites along the North and South American portion of the geosynchronous arc at 2° intervals.

Some of the techniques for spacing satellites more closely together depend on how signal polarization and satellite power levels vary between the communication spacecraft. If a signal from one satellite is relayed with its polarization opposite to that of a signal from an adjacent satellite, then interference between the same channels from these vehicles will be minimized. Alternating polarization between adjacent satellites therefore allows the spacing between them to be reduced. This technique is based on the property called "cross-polarization discrimination." This is the ability an earth station feedhorn has of detecting opposite polarity signals occupying the same frequency band at power levels 50% to 200% less than such signals having the same polarization. This method of using cross-polarization discrimination by interweaving polarization between satellites dramatically reduces interference due to closer satellite spacing.

Satellite spacing can also be reduced by locating spacecraft next to others which relay different types of communications. This method is based on the fact that power levels from communications satellites depend upon the type of information relayed. For example, while one transponder can relay hundreds of audio messages or one television broadcast, another circuit that carries two television broadcasts or one television broadcast plus a number of audio subcarriers may use higher-power microwaves. Stronger signals are more resistant to interference but are more capable of causing interference with similar signals from adjacent satellites. Therefore, if adjacent orbital positions are allocated to satellites that relay different types of information and that have different power levels interference can be reduced.

Many industry comments on the FCC request for responses to the possible change to 2 degree spacing explored such techniques. The National Cable Television Association declared it could accept 2 degree spacing if polarization interweaving was incorporated and if high-power satellites were spaced a minimum of 3 degrees apart. They claimed at that time that without these measures the fleet of 15-foot antennas used to receive satellite broadcasts would have to be replaced with antennas having at least a 23-foot-diameter at a cost to the consumer of at least two million dollars. A NASA official and a Comsat spokesperson felt that an FCC ruling would be meaningless without Canadian and South American cooperation. Most industry respondents argued for a "go-slow," careful regulatory approach, while some expressed serious doubts about the technical feasibility of 2 degree spacing. However, not all respondents expressed concern. In fact, Satellite Business Systems requested permission in April of 1982 to launch its fourth and fifth 14/12 gigahertz satellites in 1984 and 1986, respectively, so it would space its five vehicles in 2 degree intervals from 92° W to 100° W in the geosynchronous arc.

In April, 1983 the FCC finally approved 2 degree orbital spacing for satellites operating in both the C- and Ku-bands. This decision, is being implemented gradually by initially reducing spacing to 3 degrees. In time it will eventually double the capacity of the orbital arc.

Higher Frequency Microwave Transmissions

The trend towards the use of progressively higher radio wave frequencies has characterized the history of modern communications. The earlier commercial geosynchronous satellites operated in the 6/4 gigahertz portion of the C-band allocated by the FCC (see Table 7-1). Ku-band transmissions were experimentally investigated in the late 1970's by the use of the Japanese/Ford Aerospace, Sakura CS series and the Canadian Anik B series of satellites. NASA, in concert with the Public Service Satellite Consortium and the Canadian government, also participated in Ku-band experiments with one of the Anik satellites (also named the Communications Technology Satellite, CTS). Commercial broadcasts in the Ku-Band began in Canada in September of 1979, while the Ku-band satellite SBS-1 was the first commercially demonstrated in the United States in August of 1981. Today, numerous American Ku-band communication vehicles for use in all forms of communications including Ku-band television have either been launched or are in planning or construction stages.

TABLE 7-1. MICROWAVE FREQUENCY ALLOCATIONS.	
Band Name	Frequency Range
L-band	0.39 to 1.55
S-band	1.55 to 5.2
C-band	3.70 to 6.20
X-band	5.20 to 10.9
Ku-band	15.35 to 17.25
K-band	10.9 to 36.0
Ka-band	33.0 to 36.0
Millimeter	30.0 to 300.0

The employment of higher frequency microwave bands for communicating via satellite has some distinct advantages. Since antenna gains increase with frequency, smaller receiving dishes can be used. However, perhaps the most important reason for using Ku-band transmissions is the opportunity to increase the number of available communication circuits. Ku-band and C-band relays are invisible to each other, so both signals can be sent to and from two co-located satellites with absolutely no interference between them. Today even higher frequency Ka-band transmissions are under experimental investigation and are paving the way for another quantum jump in the number of available transponder circuits.

Co-Location of Satellites

Satellites can be located at the same position in the geosynchronous arc if they operate in different frequency bands or if the

TABLE 7-2. ASSIGNED USES OF MICROWAVE FREQUENCIES.

Frequency Band (GHz)	Bandwidth (MHz)	Mode	User
2.11—2.13	29		USA common carrier
2.16—2.18	20		
3.70—4.20	500		
5.925—6.425	500		
10.7—11.7	1000		
3.70—4.20	550	Downlink	Commercial satellites
5.925—6.425	500	Uplink	
7.25—7.75	500	Downlink	USA government and
7.90—8.40	500	Uplink	military satellites
10.95—11.2	2500	Downlink	Mobile radio, terrestrial
11.45—11.7	250	Downlink	microwave common carrier,
11.7—12.2	500	Downlink	and satellite links
14.0—4.5	500	Uplink	
17.7—19.7	2000	Uplink	Allocated but not yet
19.7—21.2	1500	Uplink	used—requiring more
27.5—29.5	2000	Downlink	development
29.5—31.0	1500	Downlink	
0.620—0.790			DBS Satellites. The first
2.50—2.60			three bands in use today—
11.7—12.5			last three bands under
22.5—23.0			experimental development
41.0—43.0			
84.0—86.0			

receiving and transmitting antennas of each are pointed in widely different directions. For example, in 1981 the 135° W location was populated by five vehicles dedicated to three different services: broadcasting, defense, and meteorology (see Table 7-3). Of the three defense satellites that operate in the 8/7 gigahertz band, one was a replacement for a similar vehicle, while the third had a very different footprint pattern from the other two. A South American country could have also co-located a 6/4 gigahertz satellite at the same 135° W position if its footprint were very different from the American spacecraft and if the satellite receiving antennas operated in different frequency ranges.

In the near future, clusters of non-interfering, co-located satellites will be relayed information by only one uplink facility. In addition, all these vehicles could conserve resources by sharing a single platform, positioning system, and power supply. Such a rigid plat-

TABLE 7-3. SOME CO-LOCATED OR CLOSELY SPACED SATELLITES

Satellite Description (°W)	Orbital Location	Frequency Band (GHz)
Co-Located Satellites:		
Defense Satellite Communication-DSCS11-F11 System—Eastern Pacific	13	8/7
Defense—NATO111-B—Eastern Ridge	135	8/7
Meteorological Surveys—GOES/SMS	135	2.1/1.7
Broadcast TV—Satcom I	135	6/4
Defense—DSCS11-F14—Eastern Pacific	135	8/7
Closely Spaced Satellites:		
Broadcast TV—Satcom III	132	6/4
SPC F1	132	6/4
SPC H3	132	6/4
Defense—DSCS11-F9	130	8/7
Broadcast TV—Satcom V	140	6/4
SPC F4	140	6/4

form to which all the vehicles would be attached would probably also improve the stability of all the satellite receiving and transmitting antennas and therefore enhance overall performance.

Intersatellite Communication

Presently, television or other forms of high-density communication are relayed across the Pacific or Atlantic Oceans either by transoceanic cable links or by an Intelsat satellite. Fiber-optic transoceanic cables and satellites are in direct competition and the capability of both is immense. For example, newly developed fiber-optic transoceanic cables, one of which can carry 4000 telephone conversations or more than 10 television broadcasts, are presently being designed and will be installed in the near future. But the Intelsat V satellite now positioned over the Atlantic Ocean can either relay 40 high-quality color TV broadcasts, transmit 2 million telex channels, manage more than 13,000 two-way phone calls, or send 5 trillion digital bits of information in an hour, which is equivalent to transmitting the entire Encyclopedia Britannica in 6 minutes.

Satellite circuits can be used even more effectively if the spacecraft are directly linked by microwave or laser beams. Such intersatellite communication works very well in outer space because there is no atmosphere to weaken signals bouncing between satellites. In fact, the FCC and the International Telecommunications Union have already allocated the frequency bands for intersatellite communications. Using such intersatellite relays could make communication between very distant countries less expensive by eliminating an intermediate link (see Figure 7-1). For example, a message uplinked in San Francisco could be received by a satellite over the American arc, relayed directly to another positioned over the Mediterranean, and subsequently downlinked to Saudi Arabia. The alternative would be to use an intermediate Intelsat satellite or a submarine cable, which would probably increase the cost of communication.

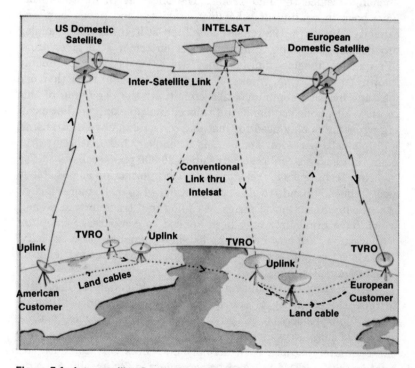

Figure 7-1. Intersatellite Communication. A direct link between an American and a European domestic satellite would greatly simplify long-range communication.

Switching Between Multiple Beams

Satellites that switch between antennas aimed towards different geographical areas on command are capable of selecting from among predefined footprints and allow the available bandwidth to be used more efficiently. Two or more different messages on the same frequency band can then be sent to two or more regions. This type of "frequency reuse" scheme also uses the available power more effectively by restricting downlink antenna beams to limited areas. The European Space Agency's L-Sat platform satellite will have this capability. The same objective can be accomplished by satellites having one or more moveable antennas which can target any selected, local areas.

Efficient Use of Bandwidth

Until recently, television and radio broadcasts were always relayed in analog forms. This broadcast mode has remained nearly unchanged since the mid 1960's. The only design improvements have been in equipment reliability, stability, and power handling capacity. In early 1980 most satellites still transmitted analog broadcasts, using one transponder per customer at a given time. "Frequency division multiplexing," or as it is more commonly known, "multiplexing," makes it possible to transmit more than one message by each communication circuit. At the beginning of this century telephone companies introduced multiplexing so that existing wires previously used for just one conversation would transmit multiple calls per wire. Today, this technology has matured to the point that a single cable can carry up to 20,000 phone calls.

New techniques for relaying as much information as possible in each frequency band are also being developed to make more efficient use of satellites. Digital forms of coding and transmitting signals which have previously only been applied to computer-to-computer conversations are now increasing satellite communication capacity. These new techniques are also relatively immune to terrestrial interference and noise.

One of the more promising digital communication methods used to raise the efficiency of transponder circuits is termed "time division multiple access" (TDMA). TDMA is comparable to time-sharing of computers in which many parties can simultaneously use one computer because each actually only uses this computer for brief intervals. TDMA use of transponders is technically more demanding than FDMA because all the earth stations served must be carefully synchronized as they must be sequentially turned on and off. However, TDMA reduces interference between signals being relayed by the same transponder and allows more than twice the information to be transmitted as compared with FDMA relays. For example, the simultaneous transmission and reception of alternating pictures, a method developed by CBS, can relay two television broadcasts per transponder using TDMA. Furthermore, Intelsat has recommended that all its member nations use a speech processing technique, Digital Speech Interpolation, which employs TDMA transmission. This permits four telephone calls to be relayed over the same bandwidth that previously could transmit only one conversation.

Other digital techniques for using bandwidth as efficiently as possible are evolving to relay information such as high definition television (HDTV). HDTV pictures approach the quality of films because 1125 lines (the number chosen for use is one of the experimental HDTV systems) instead of 525 are "painted onto" the face of a television set. The required bandwidth of 150 MHz, which would be necessary with conventional analog relays of HDTV, might be reduced to 50 megahertz by the use of digital coding and transmission techniques. Digital methods are also making inroads in all other forms of satellite communication. For instance, ABC, CBS, and NBC are beginning the process to ultimately relay all their network radio stations via digital satellite transmissions.

Teleconferencing and the Office of the Future

Since the day that Alexander Graham Bell had the first telephone conversation, which marked the occasion of the first teleconference, the "let your fingers do the walking" method has been the least expensive and most prevalent communication tool. The time and energy saved by the telephone is incalculable. This saving is now taking other forms. As television, computers, fiber-optic cables, and satellite relays have been evolving, they are bridging the gap between broadcast television and teleconferencing, as well as between the office and the home. And so a new era in communications has been born.

Teleconferencing methods fall into three broad categories. Audio conferencing is communication by telephone or by any other voice communication device. The simplest audio conference is a two-way telephone call. However, modern switching technology allows simultaneous connection of as many phones as necessary. Further, there are many technical options, such as voice-actuated speaker phones, one of which can be placed on a conference table to receive and send conversations. More sophisticated "studio-quality" devices are also available.

Computer conferencing is communication over telephone lines between computer terminals. Any number of terminals can be inter-connected. And all the computer operators need not be in attendance simultaneously because each computer terminal can "memorize" any incoming data for later digestion. This form of tele-conferencing has additional flexibility since participants can be polled on any specific point and the memorized data can be thoroughly analyzed.

Video teleconferencing, also labelled video conferencing, communicate via pictures or full-motion television images in addition to audio messages. An explosive growth in video conferencing is underway because the necessary high-frequency, broadband technologies including fiber-optic and coaxial cables, line-of-sight microwave relays, and most significantly, the satellite are now available. Both audio and video conferencing have enormous potential since 96% of all residences and offices in the United States have telephones linked to a common world-wide network.

Video conferencing can be categorized as either full or inter-active teleconferences. The full video conference involves all parties in two-way transmission of audio and video messages. This can be illustrated by the PBS network program, the MacNeil/Lehrer Report, in which participants in New York and Washington can see each other on large-screen televisions. In an interactive video conference, messages are relayed from a single uplink location to multiple receiving points. Interactive teleconferencing is used more frequently when many participants are involved because the number of uplink sites is most often limited. Any questions are usually relayed back to the central focus by telephone.

Video conferences should ideally be designed so professional quality communications are possible. A meeting can be held in a television studio leased from a local station, a hotel conference room or a corporate board room, all of which can be equipped with remote-control cameras and microphones when simplicity or privacy are necessary. Arrangements may be made for a variety of special features such as facsimile transmission of documents, slide or movie presentations, or "electronic blackboards" that transmit writing to a receiving screen. Given the requirements for establishing a noise-free communication link and for well-organized scheduling, the complexity may parallel that of a well-produced network television show.

The needs of a particular meeting must be assessed to determine whether full-motion, color video or the much less costly freeze-frame or slow-scan video is required. Slow-scan technology can relay a television picture every 10 to 90 seconds over conventional telephone lines. This is in marked contrast to transmitting a full-motion video signal in which a snapshot is "painted" on the television screen 30 times every second. The more leisurely pace of the slow-scan method permits use of a much narrower bandwidth than that of full-motion television. Slow-scan video conferencing can be perfectly adequate for situations such as a medical education seminar in which just a voice and descriptive charts, photographs and X-rays need to be communicated.

In many instances, teleconferencing can improve the quality and lower the costs of communicating between distant parties. Meetings "by satellite" can be held between widely separated parties at short notice. Travel time and expense are drastically reduced, especially if large numbers of participants are involved. For example, a 1981 Ford Motor Company teleconference brought 17,000 dealers in 38 locations "together" for a simultaneous introduction of the 1982 new car models. Teleconferencing is also finding important applications in medical education, in the exchange of medical information and even in the diagnoses of illnesses. In fact, the first live international overseas television message between the United States and Europe via the Intelsat, Early Bird satellite on June 28, 1965, was of an open-heart surgery. The earliest uses of audio conferencing over 40 years ago were also in the health care field.

The broadest application of video conferencing has been in the private business sector, resulting in significant improvements in productivity and reductions in cost. Although businesses place more importance on accurate, rapid decisions than ever before, office productivity in the last decade rose only 4% while industrial productivity rose 90%. Teleconferencing can increase office productivity by improving the flow of information and by reducing nonproductive time spent in travel. For example, M/A Com, a high-tech conglomerate with subsidiaries scattered throughout the United States, have held 15 video conferences each week on the average. Recent surveys indicate that video conferencing could replace 5% to 15% of all business meetings that require significant amounts of travel. However, many meetings of a stressful nature, such as contract negotiations, require face-to-face contact. Nevertheless, video conferencing is much more effective than audio conferencing when visual input is necessary to "get a feel" for the substance of the information being communicated.

The growth of business teleconferencing has an interesting

sidenote. As escalating energy costs have increased travel costs, the incentive to teleconference has risen and subsequently hotels and motels have experienced lower bookings. In order to offset this decrease in revenue, some large national and international hotel chains are establishing teleconferencing studios. For example, Intercontinental Hotels, owner of more than 100 hotels in 50 countries, in conjunction with Comsat General, has installed video conferencing facilities and equipment.

The popularity of teleconferencing has some negative aspects. Some experts have projected that the rapid growth of teleconferencing will strain the transponder market by requiring as many as 3500 satellite circuits by the end of this century. It has also been suggested that the rapid move towards teleconferencing may not be the best allocation of scarce transponder space. Another criticism occasionally expressed is that many executives, business people, educators and professionals would simple rather enjoy the pleasures of travelling.

It is interesting to note that some who have participated in teleconferences relayed to wide audiences have occasionally had a rather unusual experience. Following wide exposure via a video conference they have been touched by "the Hollywood effect," i.e. being well known by strangers.

Teleconferencing technologies are a key element in the office of the future, especially as more efficient uses of the available bandwidth are being developed and as costs are subsequently falling. For example, in early 1982 an RCA scientist invented a technique for relaying 2-way full-motion video conferencing over a single transponder instead of over two as was previously required. Signal scrambling techniques are also being perfected to ensure privacy when transmitting information such as sensitive business plans. Sophisticated remote-controlled TV cameras, audio monitors and special service equipment for allowing private meetings are now commercially available.

These and other related developments today allow offices and even homes to be linked using telephones, televisions or computers whose memories can store virtually whole libraries. It is clear that teleconferencing has an important role to play in our expanding communication system.

Private Business Networks

Today, many businesses are introducing low-priced, high-performance equipment for communicating voice, data and video information. New markets such as home satellite TVROs and DBS which were virtually non-existent ten years ago, have been created by these technical innovations. Satellite communication has made costs much less sensitive to the distance between the sender and the receiver than they were just 10 or 20 years ago when land-based methods were necessary. As a result, the distinction among local, regional, domestic and international markets has blurred and new business opportunities have emerged.

The telecommunication web is being woven by a host of technologies. Cities can be laced with broadband networks of coaxial and fiber-optic cables connecting homes and businesses. These threads can also be connected to line-of-sight microwave relays, satellite uplinks or satellite receiving stations. Such networks can relay any type of information desired to any type of reception equipment such as computers, televisions, telephones or facsimile devices. Any of these can be interconnected in any conceivable way.

Perhaps the most visible new businesses emerging as a result of these flexible communication technologies are local and national telephone companies. In the early 1980s AT&T had experienced serious competition in the long distance phone business from MCI and Southern Pacific Corporation. Today firms such as the American Satellite Corporation, Satellite Business Systems, M/A Com and GTE Communications, which provide similar services to those offered by the more established common carriers, are selling access to their all-digital, high speed communication networks. Satellite Business Systems, for instance, operates its own fleet of Ku-band satellites which transmit digitally encoded TDMA signals carrying voice, data and video images. Some larger corporations are establishing privately owned networks for relaying in-house, corporate information. An example is Citicorp, the large New York-based bank, which spent $60 million on telecommunication in 1980. It has substantially cut costs in linking its eight hundred offices by establishing its own network via two Citicorp-owned transponders aboard the Westar spacecraft.

The most convincing sign of the continuing growth in this business is the tremendous amount of money invested both on the earth and in space. As a result, the number of satellites and earth receiving stations has risen dramatically. Over 40 satellites have been launched by American interests since 1974 while over 3 billion cumulative dollars have been invested to date in the U.S. earth and space segments. The pace of this growth will almost certainly continue to quicken.

8

CAREERS AND OPPORTUNITIES

The rapid growth of satellite communication has created enormous opportunities for enterprising people. New businesses have emerged and many new jobs have been created. The first step in participating in this exciting field is to understand its technology and its far-reaching effects on many parts of our society. Then the choice of career paths can be more clearly seen.

Today, the most pressing demand is for communication technicians and engineers to design, build, and operate satellites, earth stations, and all other inter-connecting equipment. In particular, the home satellite television business seems to have a chronic shortage of top quality installers and service people. The level of education and experience varies widely from job to job. For example, highly skilled electrical engineers are needed to design sophisticated electronic equipment. Technicians who can be trained in a matter of months are required for installing earth stations or cable lines. Television networks and other users of satellite communication rely on experienced personnel to oversee the operation of their equipment and to be available to correct any unexpected problems.

There are also many opportunities for trained sales personnel in satellite communications. Readers of this book will have an advantage over many less informed but successful salespeople already working in this field. The information presented here will answer many questions often posed by perspective customers.

Each component of the satellite television industry has a need for talented personnel. For example, satellite master antenna television alone has a vast market waiting to be tapped. Although competition for SMATV subscribers is becoming fierce in some regions of the country, there will always be a need for competent and aggressive salespeople. The home TVRO market is also experiencing explosive growth. Similar marketing techniques can be used to sell home

TVRO systems, low noise amplifiers, cable TV subscriptions, or transponder time on-board satellites.

Competition for many more general positions in the satellite communication business is greater than that for technical or sales experts. For example, financial officers, accounting specialists or personnel recruiters need to have much less special training in satellite technologies. Similarly, television actors or camera operators face the similar stiff competition in any branch of the broadcast business. However, those who do understand the emerging businesses may have an advantage in being able to act more quickly in responding to new career opportunities.

Many have chosen the high-risk/high-return route in attempting to start their own businesses. There have been some notable success stories in recent years among those who have caught the entrepreneurial fire and have launched either manufacturing, distribution, installation or service businesses. And many excellent opportunities still exist for those who can see a small distance into the future.

THE DECIBEL NOTATION

Decibels (abbreivated as dB) are used to express the relative values of two signals. The logarithnic scale is used to compress large differences in numbers to a more manageable range. Decibels are defined by the following equation:

Decibel difference = 10 logarithm (signal A/signal B)

For example, if signal A is 1000 watts and signal B is 10 watts, then signal A is 20 dB stronger than signal B because:

$$
\begin{aligned}
\text{Decibel difference in power} \quad &= \quad 10 \log (1000/10) \\
&= \quad 10 \times 2 \\
&= \quad 20 \text{ dB}
\end{aligned}
$$

Therefore, if an amplifier received a signal of 10 watts and increased its strength by a factor of 100 to 1000 watts, it would have a gain of 20 dB. Similarly, if a 10 watt signal was increased by a factor of 1,000,000 to 10 million watts, the gain would be 60 dB.

Decibels are also expressed relative to a reference value such as watts or milliwatts. The abbreviations dBw and dBm mean the relative increase in power relative to one watt and to one milliwatt, respectively. For example, 20 dBm means a power of 100 milliwatts while 60 dBw means a power of 1 million watts.

GLOSSARY OF TERMS

Alignment   The process of fine tuning a dish or an electronic circuit to maximize its sensitivity and signal receiving capability.

Amplifier   A device used to increase the power of a signal.

Antenna   A device that collects and focuses electromagnetic energy. This process results in an energy gain, which is proportional to surface area for a microwave dish.

Aperture   The area of a parabolic antenna.

Attenuator   The passive device which reduces the power of a signal.

Audio Subcarrier   The carrier wave that transmits audio information between 5 and 8.5 MHz on a satellite broadcast.

Automatic Frequency Control (AFC)   A circuit that locks onto a chosen frequency and will not drift away from that frequency.

Automatic Gain Control (AGC)   A circuit that locks the gain onto a fixed value and thus compensates for varying input signal levels keeping the output constant.

Azimuth-Elevation (Az-El) Mount   An antenna mount which tracks satellites by moving in two directions: the azimuth in the horizontal plane; and elevation up from the horizon.

Azimuth   Degrees of rotation clockwise from true north.

Bandpass Filter   A circuit or device that allows only a specified range of frequencies to pass through a circuit.

Bandwidth   The frequency range allowed to pass through any circuit.

Baseband   The pure audio and video signal without a carrier wave. Satellite signals have audio baseband information from near zero frequency to 3400 Hertz. Video baseband is from zero to 4.2 MHz.

Beamwidth   A measure used to describe the width of vision of an antenna. It is measured between the 3 dB half power points in angles.

Block Downconversion   The process of lowering the entire satellite band of frequencies in one step to some intermediate range to be processed inside the video reciever. Multiple receivers can independantly select channels by processing this block of signals.

BNC Connector   A weatherproof twist lock coax connector used on some brands of satellite receivers and standard on commercial video equipment.

Boresight   That direction along the axis of either a transmitting or a receiving antenna.

Buttonhook Feed   A rod shaped like a question mark supporting the feedhorn and LNA. In commercial dishes a buttonhook feed often is a hollow waveguide that directs signals from the feedhorn to an LNA behind the dish.

CATV   An abbreviation for Community Antenna Television used to describe cable TV.

C-Band   The 3.7 to 4.2 GHz band of frequencies at which some broadcast satellites operate.

Carrier   A single frequency radio signal that is modulated to carry information.

Carrier-to-Noise Ratio (C/N)   The ratio of the received carrier power to the noise power in a given bandwidth. The C/N is an indicator of how well an earth station will perform in a particular location and is calculated from satellite power levels, antenna gain and the combined antenna and LNA noise temperature.

Carrier-to-Noise Ratio (G/kT)   The C/N expressed in decibels per Hertz of signal bandwidth.

Cassegrain Feed System   An antenna feed design that includes a primary reflector, the dish, and a secondary reflector which redirects microwaves via a waveguide to an LNA.

Channel   A segment of bandwidth used for one communications link.

Circular Polarization Electromagnetic waves whose electric field uniformly rotates along the signal path. Broadcasts used by Intelsat and other international satellites use circular not horizontally or vertically polarized waves.

Clark Belt The circular orbital belt at 22,247 miles above the equator, named after the writer Arthur C. Clarke, in which satellites travel at the same speed as the earth's rotation. Also called the geosynchronous or geostationary orbit.

Coaxial Cable A cable for transmitting high frequency electrical signals with low loss. It is composed of an internal conducting wire surrounded by a insulating dielectric which is covered by a metal shield.

Cross Polarization Term to describe signals of the opposite polarization than the one being transmitted. Cross-polarization discrimination refers to the ability of a feed to detect one polarity and reject the opposite polarity signals.

Decibel (dB) A term that expresses the ratio of power levels used to indicate gains or losses of signals. Decibels relative to one watt or milliwatt are abbreviated as dBw and dBm, respectively.

Direct Broadcast Satellite (DBS) A term commonly used to describe satellite broadcasts directly to homes using the Ku-band, 12 to 14 GHz.

DC Power Block A device which stops the flow of DC power but permits passage of higher frequency signals.

Declination Offset Angle The adjustment angle of a polar mount between the polar axis and the plane of a satellite antenna used to aim at the geosynchronous arc.

Demodulator A device which extracts the signal from the transmitted carrier wave.

Detent Tuning Tuning onto a satellite channel by selecting a preset resistance.

Digital Describes a system or device in which information is transferred by electrical "on-off," "high-low," or "1/0" pulses instead of continuously varying signals as in an analog message.

Dish Jargon for a parabolic microwave antenna.

Dish Illumination Describes how a feedhorn "sees" the surface of a dish as well as the surrounding terrain.

Downconverter The circuit associated with a satellite receiver that lowers the high frequency signal to a lower, intermediate range. There are three distinct types of downconversion: single downconversion; dual downconversion; and block downconversion.

Downlink Antenna The antenna on-board a satellite which relays signals back to earth.

Dual Feedhorn A feedhorn which can simultaneously receive both horizontally and vertically polarized signals.

Earth Station A complete satellite receiving or transmitting station including the antenna, electronics and all associated equipment necessary to receive or transmit satellite signals.

Effective Isotropic Radiated Power (EIRP) A measure of the signal strength that a satellite transmits towards the earth below. The EIRP is highest at the center of the beam and decreases away from this boresight.

Elevation Angle The vertical angle measured from the horizon up to a targeted satellite.

FCC The Federal Communications Commission, the regulatory board which sets standards for communications within the United States.

f/D Ratio The ratio of an antenna's focal length to diameter. It describes the "depth" of a dish.

Feedhorn A device that collects microwave signals reflected from the surface of an antenna. It is mounted at the focus in all prime focus parabolic antennas.

Focal Length The distance from the reflective surface of a parabola to the point at which incoming satellite signals are focused, the focal point.

Footprint The geographic area towards which a satellite downlink antenna directs its signal. The measure of strength of this footprint is the EIRP.

Frequency The number of vibrations per second of an electromagnetic signal expressed in cycles per second or Hertz.

Gain The amount of amplification of output to input power often expressed as a multiplication factor or in decibels.

Gain-to-Noise Temperature Ratio (G/T) The figure of merit of an antenna and LNA. The higher the G/T the better the reception capabilities of an earth station.

Geostationary Orbit See Clarke Belt.

GigaHertz (GHz) Billions of cycles per second.

Global Beam A footprint pattern used by communication satellites targeting nearly 40% of the earth's surface below. Many Intelsat satellites use global beams.

Ground Noise Unwanted microwave signals generated from the warm ground and detected by a dish.

Hardline A low-loss coaxial cable that has a continuous hard metal shield instead of a conductive braid around the outer perimeter.

Heliax A thick low-loss cable used at high frequencies also known as hardline.

Inclinometer An instrument used to measure the angle of elevation to a satellite from the surface or the earth.

Intermediate Frequency (IF) A middle range frequency generated after downconversion in a satellite receiver.

INTELSAT The International Telecommunication Satellite Consortium, a body of 154 countries working towards a common goal of improved worldwide satellite communications.

Isolator A device that allows signals to pass unobstructed in one direction but which attenuates their strength in the reverse direction.

Kelvin Degrees (K) The temperature above absolute zero, the temperature at which all molecular motion stops, graduated in units the same size as degrees Celcius (°C). Absolute zero equals -273 °C or -459 °F.

Ku-Band The microwave frequency band between 11.7 and 12.2 GHz.

Latitude The measurement of a position on the surface of the earth north or south of the equator measured in degrees of angle.

Line Amplifier An amplifier in a transmission line that boosts the strength of a signal.

Local Oscillator   A device used to supply a stable single frequency to an upconverter of a downconverter. The local oscillator signal is mixed with the carrier wave to change its frequency.

Longitude   The distance east or west of the prime meridian measured in degrees.

Low Noise Amplifier (LNA)   A device that receives and amplifies the weak satellite signal reflected by an antenna via a feedhorn. LNAs have noise temperature rated in degrees Kelvin.

Low Noise Block Downconverter (LNB)   An LNA which also downconverts the whole 500 MHz satellite bandwidth at once to an intermediate frequency range.

Low Noise Converter (LNC)   An LNA and a conventional downconverter housed in one weatherproof box.

Magnetic Variation   The difference between true north and the north indication of a compass.

Master Antenna TV (MATV)   Broadcast receiving stations that use a high-quality UHF and/or VHF antennas centrally located and that relay this TV to a local apartment/condo or group-housing complex.

MegaHertz (MHz)   Millions of cycles per second.

Microwave   The frequency range from approximately 500 MHz to 30 GHz.

Modulation   A process in which a message is added to a carrier wave. This can be done by frequency or amplitude variation, known as AM or FM, respectively.

Mount   The structure that supports an earth station antenna. Polar and az-el mounts are the most common variety.

N-Connector   A low-loss coaxial cable connector used a C-band microwave frequencies.

NTSC   The National Television Standards Committee which sets standards for North American TV broadcasts.

Noise   An unwanted signal which interferes with reception of the desired information. Noise is often expressed in degrees Kelvin or in decibels.

Noise Figure   The ratio of the actual noise power generated at the input of an amplifier to that which would be generated in an ideal resistor. The lower the noise figure, the better the device.

Noise Temperature A measure of the amount of thermal noise present in a system or a device. The lower the noise temperature, the better the device.

Offset Feed A feed which is offset from the center of a reflector. This configuration does not block the antenna aperature.

PAL Phase Alternate Line. A European color TV format different from the American NTSC.

Pad An concrete base upon which a supporting pole and dish can be mounted.

Parabola The geometric shape that has the property of reflecting all signals parallel to its axis to one point, the focal point.

Polar Mount An antenna mount that permits all satellites in the geosynchronous arc to be scanned with movement of only one axis.

Polarization A characteristic of the electromagnetic wave. Four senses of polarization are used in satellite transmissions: horizontal; vertical; right-hand circular; and left-hand circular.

Prime Focus Antenna A parabolic dish having the feed/LNA assembly at the focal point in front of the antenna.

Radio Frequency The approximately 10 KHz to 100 GHz electro-magnetic band of frequencies used for man-made communication.

Satellite Receiver The indoors electronic component of an earth station which downconverts, processes and prepares satellite signals for viewing or listening.

Scrambling A method of altering a signal indentity to prevent its unauthorized reception by persons not having decoders.

Side Lobe A construct used to describe an antenna's ability to detect off-axis signals. The larger the side lobes, the more noise and interference a dish can detect.

Single Channel Per Carrier (SCPC) A satellite transmission system that employs a separate carrier for each channel. As opposed to frequency division mulitplexing that combines many channels on a single carrier.

Signal-to-Noise Ratio (S/N) The ratio of signal power to noise power in a specified bandwidth, usually expressed in decibels.

Sparklies Small black and/or white blips or dots in a television picture indicating an insufficient signal-to-noise ratio.

Spherical Antenna An antenna system using a section of a spherical reflector to focus one or more satellite signals to one or a series of focal areas.

Splitter A device that takes a signal and splits it into two or more identical but lower power signals.

Thermal Noise Random, undesired electrical signals caused by molecular motion known, more familiarly, as noise.

Threshold A minimal signal to noise input required to allow a video receiver to deliver an acceptable picture.

Transponder A microwave repeater, receiver and transmitter, in a satellite used to amplify and change frequency of an uplinked communication channel.

Trap An electronic device that attenuates a selected band of frequencies in a signal.

Upconverter A device that increases the frequency of a transmitted signal.

Uplink The earth station electronics and antenna which transmits information to a communication satellite.

Voltage Tuned Oscillator (VTO) An electronic circuit used in a satellite receiver that generates a frequency used in selecting channels.

Video Monitor A television that accepts unmodulated baseband signals to reproduce a broadcast.

REFERENCE MATERIALS

Satellite TV Guides

Channel Guide
300 East Hampden, Suite 340
Englewood, CO 80110
(303) 654-3006

On Sat
P.O. Box 2384
Shelby, NC 28151
(800) 438-2020

• Orbit
P.O. Box 1700
Hailey, ID 83333
(800) 792-5541

• Satellite Dish Magazine   725·4342   ENTARMENT Sucs
P.O Box 8
Memphis, TN 38101
(901) 521-1580   452-3924 WHOM #

•• Satellite TV Weekly
P.O. Box 308
928 Main Street
Fortuna, CA 95540
(800) 556-8787   800 345 8876

Satellite TV Guide
P.O. Box 8266
Edmonton, Alberta T6H 4P1
Canada
(403) 425-1169

Trade Publications

Coops Satellite Digest
P.O. Box 2384
Shelb, NC 28151
(800) 438-2020

Home Satellite Marketing
International Thompson Business Press
345 Park Avenue South
New York NY 10010
(212)686-7744

Home Satellite TV
Miller Magazines, Inc.
2660 East Main Street
Ventura CA 93003
(805)643-3664

Private Cable
Weisner Publishing Company
5594 S. Prince Street
Littleton, CO 80120
(303) 798-1274

Satellite Communications
Cardiff Publishing
6430 S. Yosemite Street
Englewood, CO 80111
(303) 694-1522

Satellite Business
Steve Tolin Enterprises
P.O. Box 2772
Palm Springs, CA 92263
(619)323-2000

Satellite Dealer
CommTek, Inc.
P.O. Box 1048
Hailey, ID 83333
(208) 788-4936

Satellite Retailer
P.O. Box 2384
Shelby NC 28151
(704) 482-9673

Satellite TV Opportunities
1717 East University Avenue
Oxford, MS 38655

Satellites Times
P.O. Box 2347
Shelby, NC 28151-2347

STV Magazine
501 N. Washington Street
Shelby, NC 28150

TVRO Technology
Weisner Publications
5951 South Middlefield Road
Littleton, CO 80123

TV Satellite Videoworld
Harris Publications, Inc.
1115 Broadway
New York NY 10010

INDEX